AF564416

Recent Trends in Animal Behaviour

RECENT TRENDS IN ANIMAL BEHAVIOUR

Editors :

Shakunthala Sridhara

B. Nagachaitanya

A.K. Chakravarthy

T.K. Prabhakara Shetty

New India Publishing Agency

Pitam Pura, New Delhi- 110 088

Published by
Sumit Pal Jain *for*
New India Publishing Agency
101, Vikas Surya Plaza, CU Block, L.S.C. Mkt.,
Pitam Pura, New Delhi- 110 088, (India)
Phone: 011-27341717, Fax: 011-27341616
E-mail: newindiapublishingagency@gmail.com
Web: www.bookfactoryindia.com

'Procedings of 31st Annual Conference of Ethological Sociey of India-2007'
Sponsored by Ethological Society of India

ISBN : 978-81-89422-54-7

Composed and Designed by NIPA

Dedicated

to

Late Prof. B. R. Seshachar

First President of Ethological Society of India

Preface

The lives of animals are not made up of wholly blind, innate response to the world around them. In reality, these seemingly simple responses are the exceptions and an animal has a far more complex assessment of its environment. Complexity rather than simplicity seems to be the best description of their behaviour. Some outstanding behavioural patterns of survival value are briefly outlined below:

In the two major areas of animal's life-aggression and mating, animals make complex discriminations about each other. Many animals do seem to have a very precise knowledge about the probability that they will win a fight against a given opponent. They are able to weigh each other up, assess another animal's fighting ability relative to their own and make an entirely rational decision to fight or withdraw on the basis of their assessment. Similarly subtle evaluation of cues is shown by female animals when they choose a mate. They are less impressed with the bodily displays of the males but easily detect more subtle cues in him that will guarantee the survival of her offspring.

Rodent control campaigns have distinctly showed that the rats which were cautious about what they ate than others, survived the poisoning, reproduced and likely to have passed on their cautiousness to their offspring. Even more extraordinary phenomenon is that an animal not only learns for itself and profits from its mistakes but it is also capable of learning from the experience of other animals as rats use each other as "testers" for what might be safe or poisonous food.

'Decision making' is exhibited by many animals after weighing up different influences. Sometimes priority is given to avoid danger, sometimes to get enough food but usually there is an uneasy compromise between the two in foraging and feeding behaviour. 'Co-operation' i.e. interacting with other members of species to achieve a greater benefit to all the members as a group than an individual would achieve on its own, is exhibited by many animals. The vampire bats take co-operation to new level by giving food selectivity to those bats that proved to be valuable reciprocators in the past. It looks as if they have learnt that present is not important but long-term bonds between individuals pay off in the long run.

Of late, ethology has moved into technological direction. Neuro-endocrine basis of behaviour, mechanization of data collected and use of molecular biology techniques are the latest tools to understand animal behaviour. The 31st annual conference of Ethological Society of India indicates the directions in which Indian Ethologists are involved at present. We hope the proceedings will be of great bonding effect amongst the Ethologists of India.

Bangalore
22-05-2008 *Editors*

Contents

SECTION - 2
BEHAVIOURAL ECOLOGY

SECTION - 4
FORAGING AND FEEDING BEHAVIOUR

SECTION - 5
APPLIED BEHAVIOUR

SECTION - 6: REPRODUCTIVE AND SEXUAL BEHAVIOUR

List of Contributors

Alok Bang
Centre for Ecological Sciences
Indian Institute of Science
Bangalore - 560 012

Anindita Bhadra
Centre for Ecological Sciences
Indian Institute of Science
Bangalore-560 012

Anindya Sinha
National Institute of Advanced
Studies (IISc Campus)
Bangalore-560 012

Aniruddha Mitra
Centre for Ecological Sciences
Indian Institute of Science
Bangalore - 560 012

Ashok Kumar, C. T.
Department of Agricultural
Entomology
University of Agricultural Sciences
GKVK, Bangalore - 560 065

Avinash, T.P.
Animal Behaviour and Wetland
Research Laboratory
Department of Zoology
Chirst College, Irinjalakuda
Kerala-680 125

Azad Thakur, N.S.
Division of Entomology
ICAR Research Complex for NEH
Region, Umroi Road, Umiam
Meghalaya-793 103

Babu, S.
Division of Wildlife Biology
Kerala Forest Research Institute
Peechi, Thrissur
Kerala-680 653

Balikai, R.A.
All India Co-ordinated Sorghum
Improvement Project
(UAS Dharwad)
Regional Agri. Research Station
Bijapur -586 101
Karnataka

Basavaraju, B.S.
Agricultural Research Station
UAS (B), Madenur, Hassan
Karnataka

Bharathiraja, A.
P.G. and Research Department
of Zoology, Nehru Memorial College
(Autonomous), Puthanampatti
Tiruchirappalli – 621 007
Tamil Nadu

Bhat, N.S.
Department of Apiculture
University of Agricultural Sciences
GKVK, Banglaore-560 065

Binoy, V.V.
Animal Behaviour and Wetland
Research Laboratory
Department of Zoology
Chirst, College, Irinjalakuda
Kerala-680 125

Biradar, A.P.
Department of Entomology
College of Agriculture
Bijapur-586 101, Karnataka

Chakravarthy, A. K.
Department of Agricultural
Entomology, University of
Agricultural Sciences, GKVK
Bangalore – 560 065

Chandrashekara , K.
Department of Entomology
University of Agricultural Sciences
GKVK, Bangalore-560 065

Chandrashekaran, M.K.
Evolutionary and Organismal
Biology Unit, Jawaharlal Nehru
Centre for Advanced Scientific
Research, Jakkur, P.B. No. 6436
Bangalore – 560 064

Changalarayappa
Department of Sericulture
University of Agricultural Sciences
GKVK, Bangalore-560 065

Chaya Kumari
Unit of Evolution and Genetics
Department of Studies in Zoology
University of Mysore
Manasagangotri
Mysore-570 006

Chinnaswamy, K.P.
Department of Sericulture
University of Agricultural Sciences
GKVK, Bangalore-560 065

Debjani Roy
Centre for Distance Education
North Eastern Hill University
Bijni Complex, Laitumukhrah
Shillong – 793 003
Meghalaya

Devaraj, K.
Butterfly Park, Zoo Authority of
Karnataka, Bannerghatta
Bangalore

Emmanuvel Rajan
Department of Animal Science
School of Life Sciences
Bharathidasan University
Tiruchirappalli-620 024

Fatima, S.
Department of Sericulture
University of Agricultural Sciences
GKVK, Bangalore-560 065

Gangappa, E.
Department of Genetics and Plant
Breeding, University of Agricultural
Sciences, GKVK
Bangalore – 560 065

Ganiger, P. C.
Department of Entomology
University of Agricultural Sciences
GKVK, Bangalore -560 065

Gavigowda
Department of Apiculture
University of Agricultural Sciences
GKVK, Bangalore-560 065

Geetha Bali
Vice-Chancellor
Karnataka Women's University
Bijapur

Girish, A.C.
Center for Sustainable Technology
Indian Institute of Science
Bangalore – 560 012

Gopukumar, N.
Department of Animal
Behaviour and Physiology
School of Biological Sciences
Madurai Kamaraj University
Madurai - 625 021

Guruprasad, H.
Department of Entomology
University of Agricultural Sciences
GKVK, Bangalore - 560 065

Harini, B. P.
Zoology Department
Bangalore University
Bangalore-560 056

Hema Krishna Kumar
Department of Life Sciences
National Taiwan University
Taipei 106, Taiwan

Hsiung Liang
Institute of Life Science
National Kaohsiung Normal University
Kaohsiung 804, Taiwan

Iqbal Perwez
Department of Zoology
Aligarh Muslim University
Aligarh, Uttar Pradesh

Irulan
Department of Animal
Behaviour and Physiology
School of Biological Sciences
Madurai Kamaraj University
Madurai – 625 021

Jagadish, N.
Department of Sericulture
University of Agricultural Sciences
GKVK, Bangalore-560 065

Jagginavar, S. B.
Department of Agricultural
Entomology, College of Agriculture
Bijapur-586 101
Karnataka

Jayahari, K. M.
Division of Forest Ecology and
Biodiversity Conservation
Kerala Forest Research Institute
Peechi, Kerala - 680 653

Jayappa, A.H.
Department of Agricultural
Entomology, University of
Agricultural Sciences, GKVK
Bangalore-560 065

Jayappa, J.
Department of Entomology
University of Agricultural Sciences
GKVK, Bangalore 560 065

Jayson, E. A.
Division of Wildlife Biology
Kerala Forest Research Institute
Peechi, Thrissur
Kerala-680 653

John Shiron
Animal Behaviour and Wetland
Research Laboratory
Department of Zoology
Chirst College, Irinjalakuda
Kerala-680 125

John Thomas, K.
Animal Behaviour and Wetland
Research Laboratory
Department of Zoology
Chirst College, Irinjalakuda, Kerala

Julie, E.
Division of Acarology
Department of Zoology
University of Calicut, Kerala

Juliet Vanitharani
Bat Research Laboratory
Department of Zoology
Sarah Tucker College, Tirunelveli
Tamil Nadu - 627 007

Jyothi, H.K.
Zoology Department, P.E.S College
Hanumanth Nagar, Bangalore

Kanagaraj, C.
Department of Animal Science
School of Life Sciences
Bharathidasan University
Tiruchirappalli-620 024

Kanimozhi, M.
Department of Microbiology
VPMM College for Women
Krishanakoil, Tamil Nadu

Karuppudurai, T.
Department of Animal Behaviour
and Physiology
School of Biological Sciences
Madurai Kamaraj University
Madurai - 625 021

Krupanidhi, S.
Sri Sathya Sai University
Prasanthinilayam- 515 134
Andhra Pradesh

Kumar, A.R.V.
Department of Entomology
University of Agricultural Sciences
GKVK, Bangalore-560 065

Kumar, D.
Division of Entomology
ICAR Research Complex for NEH
Region, Umroi Road, Umiam
Meghalaya-793103

Lalitha Vijayan
Salim Ali Centre for
Ornithology & Natural History
Coimbatore-641 108, Tamil Nadu

Mahadevappa
Butterfly Park
Zoo Authority of Karnataka
Bannerghatta, Bangalore

Malathi, U.S.U.
Bat Research Laboratory
Department of Zoology
Sarah Tucker College, Tirunelveli
Tamil Nadu-627 007

Mallik, B.
Department of Entomology
University of Agricultural Sciences
GKVK, Bangalore-560 065

Manjula Kumari, D.
Department of Microbiology
Bangalore University
Bangalore-560 056

Marimuthu, G.
Department of Animal
Behaviour & Physiology
School of Biological Sciences
Madurai Kamaraj University
Madurai – 625 021

Marimuthu, K.
S. P. College of Science and Arts
Madurai

Maulishree Agrahari
Centre for Ecological Sciences
Indian Institute of Science
Bangalore 560 012 and
Department of Biotechnology
PES Institute of Technology
Bangalore-560 085

Monika Sadananda
Department of Applied Zoology
Mangalore University
Mangalagangothri
Mangalore-574 199

Muniswamy Gowda, K. N.
Department of Agricultural
Entomology, College of Agriculture
V. C. Farm, Mandya, Karnataka

Muralikrishna, M.
Department of Entomology
University of Agricultural Sciences
GKVK, Bangalore-560 065

Nagaraja, C.J.
Department of Apiculture
University of Agricultural Sciences
GKVK, Bangalore-560 065

Nandihalli, B. S.
Department of Entomology
University of Agricultural Sciences
Dharwad, Karnataka

Neelanarayanan, P.
P.G. and Research Department of
Zoology, Nehru Memorial College
(Autonomous), Puthanampatti
Tiruchirappalli – 621 007
Tamil Nadu

Pal, S.K.
Katwa Bharati Bhaban
P.O.- Katwa Dist., Burdwan
West Bengal-713 130

Panjaly Ramaraj
Thiagaraja College, Madurai

Patil, D.R.
Department of Entomology
College of Agriculture Bijapur
Karnataka-586 101

Patil, S. U.
Agricultural Research Station
AICRP (Rice Entomology)
Kankanady, Mangalore-575 002
Karnataka

Prasanna Kumar, N.R.
Department of Agricultural
Entomology, University of
Agricultural Sciences, GKVK
Bangalore-560 065

Priya, L.
Centre for Ecological Sciences
Indian Institute of Science
Bangalore-560 012

Raghavendra Gadagkar
Centre for Ecological Sciences
Indian Institute of Science
Bangalore 560 012 and Evolutionary
and Organismal, Biology Unit
Jawaharlal Nehru Centre for
Advanced Scientific Research
Jakkur, Bangalore 560 064

Raghuram, H.
Department of Animal Behaviour
and Physiology, School of Biological
Sciences, Madurai Kamaraj University
Madurai- 625 021

Rajasekar, R.
Department of Animal Behaviour
and Physiology
School of Biological Sciences
Madurai Kamaraj University
Madurai-625 021

Ramani, N.
Division of Acarology
Department of Zoology
University of Calicut, Calicut
Kerala-673 635

Ramesh, N.
Department of Zoology and Wildlife
Biology, A.V.C.College, (Autonomous)
Mayiladuthurai-609 305
Tamilnadu

Ramesh, S.R.
Unit on Evolution and Genetics
Department of Studies in Zoology
University of Mysore
Manasagangotri, Mysore-570 006

Rashmi, K.
Department of Sericulture
University of Agricultural Sciences
GKVK, Bangalore-560 065

Reshma Shetty
Department of Applied Zoology
Mangalore University
Mangalagangothri
Mangalore-574 199

Revanna Revannavar
College of Horticulture
University of Agricultural Sciences
Mudigere, Karnataka

Rohini Balakrishnan
Centre for Ecological Sciences
Indian Institute of Science
Bangalore-560 012

Ruchira Sen
Centre for Ecological Sciences
Indian Institute of Science
Bangalore-560 012

Rutuja Chitra Tarak
Centre for Ecological Sciences
Indian Institute of Science
Bangalore-560 012

Saikat Chakraborty
Unit on Social Insect Biology
Centre for Ecological Sciences
Indian Institute of Science
Bangalore-560 012
Karnataka

Sakthivel, P.
P.G. and Research Department
of Zoology, Nehru Memorial
College (Autonomous)
Puthanampatti-621 007
Tiruchirappalli - Dt.
Tamil Nadu

Samina Hisamuddin Farooqi
Department of Wildlife Sciences
Aligarh Muslim University
Aligarh, U.P

Sangeetha G Kaimal
Division of Acarology
Department of Zoology
University of Calicut, Calicut
Kerala-673 635

Sannaveerappanavar, V.T.
Department of Entomology
University of Agricultural Sciences
GKVK, Bangalore –560 065

Sathish, S.
Department of Apiculture
University of Agricultural Sciences
GKVK, Banglaore-560 065

Sathyanarayana, M.C.
Department of Zoology and
Wildlife Biology
A.V.C. College (Autonomous)
Mayiladuthurai-609 305
Tamilnadu

Satish Kumar
Department of Wildlife Sciences
Aligarh Muslim University
Aligarh, U.P

Saubhik Ghosh
Centre for Ecological Sciences
Indian Institute of Science
Bangalore-560 012

Sheeja, U.M.
Division of Acarology
Department of Zoology
University of Calicut
Malappuram-673 635

Shiji, M, T.
Division of Acarology
Department of Zoology
University of Calicut
Calicut-673 635, Kerala

Shivashankar, T.
Extention Education Unit
Agricultural Research Station
Nagenahalli
Mysore

Shivayya,V.
AINP on Rodent Control
University of Agricultural Sciences
GKVK, Bangalore-560 065

Shobha, A.
Department of Microbiology
Bangalore University
Bangalore 560 056

Shylesha, A. N.
Division of Entomology
ICAR Research Complex for NEH
Region, Umroi Road, Umiam
Meghalaya-793 103

Siddapaji, C.
Department of Sericulture
University of Agricultural Sciences
GKVK, Bangalore-560065

Singh, P.K.
Department of Applied Zoology
Agra University, Agra, Uttar Pradesh

Sowmini Balakrishnan
A-19, "Sivagami", Flat 23,
Sivakamipuram
1st Street, Thiruvanmiyur
Chennai - 600 041

Sripathi, K.
Department of Animal Behaviour
and Physiology
School of Biological Sciences
Madurai Kamaraj University
Madurai-625 021

Sujata A. Deshpande
Centre for Ecological Sciences
Indian Institute of Science
Bangalore-560 012

Suma, S.
Department of Apiculture
University of Agricultural Sciences
GKVK, Bangalore-560 065

Sumana, A.
Centre for Ecological Sciences
Indian Institute of Science
Bangalore -560 012

Sunitha, N. D.
Department of Agricultural
Entomology, College of Agriculture
UAS (Dharwad), Bijapur -586 101
Karnataka

Swami Doss, D. P.
Department of Animal Behaviour and Physiology
School of Biological Sciences
Madurai Kamaraj University
Madurai-625 021

Swati Samudre
Centre for Ecological Sciences
Indian Institute of Science
Bangalore-560 012

Thippaiah, M.
Department of Entomology
University of Agricultural Sciences
Bangalore -560 065, Karnataka

Thiruchenthil Nathan, P.
Department of Animal Behaviour and Physiology
School of Biological Sciences
Madurai Kamaraj University
Madurai-625 021

Thyagaraj, N.E.
Agricultural Research Station
University of Agricultural Sciences (B)
Arsikere-573 103

Ullas Karanth, K.
Technical Director -Tigers Forever
Wildlife Conservation Society (WCS)
New York, USA

Vidhi Pareek
Centre for Ecological Sciences
Indian Institute of Science
Bangalore

Vijay Kumar, L.
Department of Agricultural Entomology, University of Agricultural Sciences, GKVK
Bangalore – 560 065

Vishwas, A.B.
Department of Apiculture
University of Agricultural Sciences, GKVK
Bangalore-560 065

Wadnerkar, D. W.
Department of Entomology
Marathawada Agricultural University, Parbhani-431 402
Maharashtra

Yahya, H.S.A.
Department of Wildlife Sciences
Aligarh Muslim University
Aligarh, U.P

Yashwant, H.M.
Department of Apiculture
University of Agricultural Sciences, GKVK
Banglaore-560 065

SECTION - 1
KEY NOTE ADDRESS AND PLENARIES

1

Biological Clocks in Bats Mice and Humans

CHANDRASHEKARAN, M.K.

ABSTRACT

Bats: A colony of 500 insectivorous bats of *Hipposideros speoris* inhabited a natural cave in Madurai (9°58′ N. lat; 78°10′ E long). At a depth of 40 m, the cave experiences a constant temperature of 27 ± 0.1°C, 95 % relative humidity and absolute and perpetual darkness. Therefore there were no day/night cues. Yet all the members of the bat colony flew out 10 to 15 min. after sunset as though they knew the time of sunset. Recorded flight and rest experiments in flight activity cages and a hand-wound Thermohygrograph drum 40 m from the cave mouth over periods of 50 days revealed that the free-flying conspecific bats were *socially* synchronizing the activity patterns of the *three* captive bats. Bats were telling other bats time.

Mice : In the field mouse, *Mus booduga* the mother conveys information of day and night to her pups by means of her presence and absence (PA) cycles. Her presence meant day and absence denoted night. We named this phenomenon *maternal entrainment.* Such *entrainment* of the time sense of the pups was possible if in every cycle the mother was presented for a minimum of 8 hr but not exceeding 16 hr. Therefore 8 hr and 16 hr describe the limits of maternal entrainment.

Humans : Thirteen human subjects (4 females and 9 males) aged 20 to 35 years were studied for their sleep/wake cycles, rectal temperature profiles, meal timings, blood pressure, short interval estimation, alertness etc. in a *social isolation facility* with fortified walls and without windows. The subjects did not have access to a wristwatch or commercial radio and TV and therefore could not have known local time. In all the sixteen experiments the subjects *underestimated* time. We also discovered and first reported that the menstrual cycle in the human female is *not* coupled to the sleep/wake cycles. The subjective sense of duration of 2 hr intervals expanded and contracted like a time systole as a function of duration of wakefulness. The rectal temperature rhythm and the sleep-wake rhythm showed vastly different periods and in some cases even became internally dissociated.

All experiments summarized here are first reports of their kind and were performed by me and my Ph.D. students in the Department of Animal Behaviour and Physiology, School of Biological Sciences, Madurai Kamaraj University between 1975 and 1996.

2

Education Scenario in India Need for New Concepts and Region Oriented Focus

DEBJANI ROY

ABSTRACT

Education begins at home, then in community and finally in the educational institutes. At present 90% of the children in the age group of 6-14 years are receiving free primary education. Education policy of the 11th plan states that there should be free and compulsory secondary education for all. To meet this goal, additional schools have to be set up to cater to the 30 million students. To teach the 30 million students atleast one million teachers would have to be trained to ensure the 1:30 teacher-student ratio. How do we plan to meet this target?

The projected target has to be met by developing innovative teaching methods in the institutes of higher learning. These methods have to be technology based, as well as community based regional approach. Emphasis has to be given to Liberal Education where there has to be a close interaction between the teacher and learner. The Open and Distance Learning is one such System which can be compared to Liberal Education–where there is flexibility in time and pace of learning for the learners. Reforms and innovations have to be brought in this system of education with utmost caution so that there is no compromise on quality and ethics, since every age has its own ethical environment.

Ethics in this context means traditions and culture of the small communities. This is more so in the North Eastern Region where there is an intimate connection between large and small communities. This region is rich in its community based values, bioresources and cultures. Need based Vocational and Professional courses have to be designed for enriching the region with human resources infused with sophisticated skills, resulting in the economic growth for the region. This method is envisaged to take care of the social challenges, traditions and cultural ethos.

Distance Education coupled with Information Communication Technology plays a pivotal role in meeting the requirements of today's knowledge society by cutting across the barriers of time and geographic isolation.

INTRODUCTION

Education is conceived as an important instrument for social and economic change and an investment for better future. It is considered as an investment for human resource development, which would ultimately contribute to economic, industrial and scientific development of the nation. It develops skills, communicates knowledge and increases the efficiency of workers who are engaged in the task of nation building in various capacities and occupations (Shaffila, 1994).

Higher education aims not only to impart and enhance specialised knowledge and skills in different areas of learning, but also develops leadership qualities in its subjects in order to take up leadership roles in various fields. These qualities include cognitive abilities and affect attributes. The cognitive abilities are of crucial significance since it would be called upon to examine critically different situations, analyze them to identify the problems in the context of community needs and take decisions through a problem solving approach. Since the leadership role has to be played in different walks of life in close association with the members of the community, certain attributes like social sensitivity, social justice, an outlook of equality and scientific attitude needs to be imbibed through higher education (Menon & Joshi, 1986).

Education or Higher Education is a life long process. Formal education ends at some point of time but learning does not end. It continues even with employment but perhaps in different form. Learning now becomes performance related. The learner has to be flexible in acquiring skills for handling varied tasks. Thus, at one point in everyone's life, technical know-how becomes important. It is not merely for the workers who need to know the skills to operate machines but it is required for all and sundry. It is for the supervisors, managers, executives, teaching faculty who need to acquire skills to grow and acquire maturity of thought and action. In other words, one can say that learning continues but the elements in the process undergo a change.

In view of the various types of global knowledge acquisition processes, the following concepts need to be adopted in order to promote excellence in Higher Education:

1. *Learner Oriented Training Programme* : Learner oriented training and development programmes need to be formulated. These programmes will be able to impart skills ranging from technical skills in order to update with new technology, interpersonal skills in order to maintain congenial relationships with all, problem solving skills to tap the logic, analytical ability and reasoning in order to understand the causes of problems and determine solutions accordingly, and ethical skills to add values to the organisation. The latter is extremely vital in as much as others as the youth enter the field of work with pecuniary goals. They tend to lose track of their values and ethics, which are the cornerstones of humanity. The other skills include creativity, entrepreneurial drive, leadership, team commitment and so on (Mansukham, 2004).

2. *Time Based Strategy Programme* : Today's competitive environment calls for learner programmes, which are sharp, agile and quick, surpassing competitors. A time-based strategy warrants an organisation to embrace change and adapt to it easily. It advocates doing things faster than competitors and flourishing in an environment of continuous change. In the Indian context, while all the Indian companies are working up to the reality of time-based competition, even the educational institutes have been introducing innovative and job-oriented courses for speedily satisfying the needs and wants of the society, resulting in substantial revenue generation for them. Considering time as an important source of competitive change, there should also be a focus on the learner, developing flexible structure, speedy working culture, ensuring quick decision-making and effective leadership (Shah, 2002).

3. *Information Communication Technology Programme* : The changing global scenario represents real challenges to the future growth of educational institutions. More and more institutions are adopting the new Information Communication Technology (ICT) as a means of achieving competitive advantage as well as to meet the upswing in demand for their services. In order to meet the demands of the learning society and a life long learner, the institutions are offering flexible learning by providing mechanisms like modularization, accreditation of prior learning, open and distance learning and collaboration between different organizations and technological devices such as CD ROM materials, www and internet based education (Cooper, 1991). Thus, the traditional education approach with its limited seats has to change its strategy and envision new approaches to instruction based on communication and technology in order to meet the challenges of "learning on demand" and "learner centered instruction". The approach should be moving from single mode delivery to multimode with major interactive components. Traditional academic institutes should encourage their full time students to take atleast one course either online or any web based course to develop the ability to learn independently (Venugopal & Srivastava, 2002).

4. *Vocational Education and Training* : A tremendous shift from cognitive learning strategies to integrated learning strategies is taking place to assist the learners in using a variety of learning tools and methods, according to the needs and changing work environment. Employers now look for individuals who are knowledgeable, trainable, more flexible and have greater capacity to adjust according to the needs of the organisation. Vocational Education focuses on 'generic skills' and development of positive attitude. It imparts better communication skill, problem solving skill and self-learning skills to enhance employability. The length of vocational courses is diverse, favouring learning. There are variations in the programme to suit the needs of the learner in terms of economic status, educational background, training needs, accessibility, interest, occupation, employment status and achievement level. The training can cater to students of general education, differently-abled persons and physically challenged persons, school drop-outs, educated unemployed youth, illiterate youth and retrenched workers (Mehrotra, 2005).

Having dwelled upon the basic concepts which needs to be introduced in the global context of higher education, the author now will focus on the ground reality, on the more 'immediate' aspects of the educational system pertaining to the North Eastern region. What are the hurdles the students face in North-East and what steps can be taken to iron out the difficulties? Are the problems so structural in nature that it is a lost cause to even contemplate changes or is there hope? Such questions will be raised, and, with the help of reliable data, certain solutions shall be put forward.

It is generally agreed that the educational system is beset with a few basic problems – the inability to include the lower segments of society in the educational framework, the disjuncture between the educational curriculum and the needs of the society, overemphasis on a certain sector or stream in the educational system and the production of unemployable graduates from institutions that remain untouched by the changes sweeping our society (Table 1). Of late, it has been noticed that there is a general feeling of despondency about the higher education system, especially at a time when the emergence of the 'knowledge economy' is putting India on the global map (Abraham, 2000). This despondency can be attributed to a few factors. We are witness to a steady growth in enrolment for higher education without commensurate growth in the relevant sectors that should absorb the products of such a system. Such a situation has been compounded by the fact that irrational shifts have taken place in recent years in government thinking and orientation as regards higher education, and consequently, the level of its commitment to higher education. Globalization has also been a culprit. It has unleashed new forces that determine the types of institutions that are encouraged (at the cost of others) and allowing the entry of new players into the arena who are intent on furthering their self interest.

While these may be listed as the general malaise affecting the educational system in India, as regards the North-East, a few specific ones may be identified:

a) The North East is a witness to low student enrolment. All North-Eastern states barring Assam lag behind the rest of India, especially the South, East and Western parts (Table 2). The severity of this problem can be gauged when one puts the enrolment figure for India in the global context–the gross enrolment in educational institutions in India is only 7% (for higher education), which is lower than the average for Asia as a whole (11%) and way behind the figure for developed countries (Table 3) (Altbach, 2005).

b) While the rate of literacy is encouraging, there are serious problems regarding the quality of the educational curriculum, especially pertaining to science and mathematics.

c) Due to difficult terrain and chronic law and order problems, education does not reach a large part of the population of the North East.

Table 1. Subject-Wise Number of Colleges in the Country (2001-2002)

Subject	Number
1. Arts, Science, Commerce & Oriental Learning Colleges	11,128
2. Teachers Training	784
3. Engineering/Technology/Architecture	1,077
4. Medical	1,253
5. Allopathy	262
6. Ayurveda	189
7. Homeopathy	141
8. Unani/Tibbia	29
9. Dental	142
10. Nursing	122
11. Pharmacy	241
12. Physiotherapy	120
13. Naturotherapy	5
14. Public Health	2
15. Agriculture	106
16. Veterinary/Animal Science	50
17. Law	368
18. Others*	671
Total	15,437

*Others includes Colleges exclusively for Library Science, Physical Education/Yoga, Music/Fine Arts, Social Work, Journalism/Mass Communication etc. & Colleges for which type is not available

Source: University Grants Commission

Note: The data was valid as of January 1, 2002

Table 2. State-Wise Student Enrolment (2002-2003)

S. No.	State/UT	Total Enrolment	Women Enrolment	Women %
1.	Andhra Pradesh	7,51,476	2,95,330	39.30
2.	Arunachal Pradesh	5,439	1,577	28.99
3.	Assam	2,01,132	81,458	40.50
4.	Bihar	4,94,035	1,17,628	23.81
5.	Chhattisgarh	1,52,813	56,082	36.70
6.	Delhi	1,72,218	79,220	46.00
7.	Goa	19,437	11,371	58.50
8.	Gujarat	5,19,021	2,29,483	44.21
9.	Haryana	2,49,118	1,02,138	41.00
10.	Himachal Pradesh	89,914	38,383	42.69
11.	Jammu & Kashmir	59,599	28,071	47.10
12.	Jharkhand	1,97,349	59,994	30.40
13.	Karnataka	5,57,645	2,27,846	40.86
14.	Kerala	2,51,157	1,50,682	60.00
15.	Madhya Pradesh	4,74,813	1,76,630	37.20
16.	Maharashtra	12,58,195	5,15,868	41.00
17.	Manipur	36,379	16,115	44.30
18.	Meghalaya	29,165	13,970	47.90
19.	Mizoram	11,240	5,249	46.70
20.	Nagaland	12,339	4,788	38.80
21.	Orissa	3,45,113	1,23,171	35.69
22.	Punjab	2,43,743	1,28,411	52.68
23.	Rajasthan	3,63,172	1,17,416	32.33
24.	Sikkim	4,103	1,654	40.31
25.	Tamil Nadu	7,13,045	3,21,583	45.10
26.	Tripura	20,145	8,219	40.80
27.	Uttar Pradesh	11,77,810	4,52,279	38.40
28.	Uttaranchal	1,15,318	46,127	40.00
29.	West Bengal	6,48,178	2,55,382	39.40
30.	A & N Islands	2,015	1,164	57.77
31.	Chandigarh	33,942	18,838	55.50
32.	Daman & Diu	587	275	46.85
33.	Pondicherry	18,178	9,562	52.60
	Total	**92,27,833**	**36,95,964**	**40.05**

Source: University Grants Commission, New Delhi, 2003

d) It is strongly felt that the educational curriculum needs to be re-oriented as it is in variance with the broad cultural ethos of the North-East. Not enough is incorporated to hone and augment the natural qualities that students in the region develop as a result of their exposure to their society and environment.

e) The general education system provides students with low employment opportunities and when vocational education is offered to mitigate the low employment scenario, the latter is accorded a low priority, it being the first casualty of cuts in funding and general austerity measures.

f) The prioritization of a largely alien model of education has resulted in sacrificing the holistic view that a more indigenous educational system equips the person with. Stress is laid on the parts as opposed to the whole, where students are forced to specialize in a certain 'stream' after completing their 10th level exams. While such measures can and do ensure high marks in the competitive exams (which have become the Holy Grail to judge a student's calibre), it is the student's intellect that is eventually confined within the narrow walls of 'specializations'. This leads to the development of a degree of insensitivity on the part of the students, who are incapable of associating and identifying themselves with the overall system, thereby leading to socio-economic disparities, which are given a fillip by globalization.

Table 3. Gross Enrolment Ratios in Higher Education in Developed and Developing Countries (Altbach, 2005)

World total (1997)	17.40	USA (1995)	92.00
Developed Countries (1997)	51.60	Canada (1995)	88.00
Developing Countries (1997)	10.30	Australia (1997)	80.00
Less Developed Countries (1997)	03.20	Korea (1997)	68.00
Asia (1997)	11.10	UK (1996)	52.00
		Japan (2002)	45.00*
		India (1997)	07.00
		Africa (1997)	06.90
		China (1997)	06.00

A malaise that one witnesses along the length and breadth of the country, the North East being no exception is the fact that the student-teacher relationship is changing at a rapid rate. Education has become a kind of a commodity, where the students are no longer the recipients of knowledge, but customers. Seldom do students approach an institution with the express desire to benefit from a particular teacher's expounding of knowledge. Rather, they approach institutions which are 'prestigious', institutions that guarantee them a job. The brand name of the institute has a market value determined according to the number of rank holders it can attract. In turn, the value of a teacher is determined by his or her ability to 'package' information that can be retrieved and applied by the students in competitive exams. Teachers who can 'dumb down' their subject matter are highly valued in the education market. The teacher is nothing but a paid worker in the assembly line of 'packaging' knowledge. In turn, the student is a customer with the specific demand and willing to pay a price for it. In the age of globalization, even education is being commodified at a stupendous rate, in turn affecting social dynamics such as those between the teacher and the taught. Such trends can only lead to the dropping out of students who do not have the means to pay for their education and add on to the generation of educated unemployed youth.

In order to make a real difference in the lives of individuals, in the north eastern (NE) region of India, the Government of India, has set up the North Eastern Regional Education Council (NEREC) in July 2005 for revamping of education in the NE region, of which the author is the Coordinator. NEREC has set up five sub-committees to go into specific issues viz., infrastructure in educational institutions; IT-connectivity of educational institutions; upgrading quality of teaching of science and mathematics; meaningful education; and vocationalisation of education; organized seminars in practically all the universities of the region to identify the issues and suggest the solutions thereof. NEREC has formulated a road map for rejuvenating the educational system in the North east in the context of the socio-economic milieu relevant to the needs of the people of the region. The plan of action suggested by NEREC is as follows:

A. *Develop Centers of Excellence:* Instead of spreading the resources too thin by having all courses at all the places, it would be far better for the educational institutes to carefully select subjects, keeping in view their relevance to the local ethos of the area and the community, and develop Centers of Excellence around those subjects. These could then be shared by all with the help of information technology (IT).

B. *Cluster approach for quality up-gradation*: A system of organizing Colleges and Schools around a centrally located Mother Institution should be taken up so that facilities and training potential of such focal institutions, appropriately strengthened, could be shared by students and teachers of feeder institutions through periodic visits and interaction. Pooling of the limited resources is perhaps the only practical and affordable way of providing better inputs and facilities to students and training to teachers in widely scattered institutions. The general idea is to start with improving at one existing college per district and one higher secondary school per block as institutes of excellence. These should emerge as hub for networking under the umbrella of the university and serve as Mother Institution for feeder institutions in their hinterland by sharing with them their upgraded facilities like libraries, laboratories and residential hostels, games and sports and information communication technology (ICT) facilities. Students from the Mother Institution should be sent to centers of higher learning periodically and thus gain regular access to modern educational facilities. Teachers in these Mother Institutions could be trained as trainers/resource persons for training and orientation of teachers in the feeder institutions. The ultimate aim is to reach, in stages, the most isolated primary school in the remotest village of North-east region.

C. *Improving Teaching by Intensive Teacher Training Programme* : Most children in their formative years look upon their teachers as their role model. It is in the primary school that a child needs to have a very good trained teacher who would be able to enthuse the children to exercise their innate qualities, develop their interest, teach them to have a focussed learning interest and avoid rote learning. This can only happen if the schools have trained teachers and have the proper teacher-student ratio. The data of the 7th All India School Education Survey in 2002 conducted by NCERT brings forth the urgency of teachers training programme (Table 4).

Looking at the urgent need for reforms in the education system, the National Curriculum Framework (NCF) in 2003 enunciated 6 principles:

Table 4. Number of Untrained Teachers in North-eastern states of India

(i) Secondary Stage

S. No	State	Total Teachers	Untrained Teachers	% of Untrained
1	Arunachal Pradesh	1269	517	40.74
2	Assam	51194	35816	69.96
3	Manipur	5018	2911	58.01
4	Meghalaya	4845	2819	58.18
5	Mizoram	2640	1017	38.52
6	Nagaland	2046	1494	73.02
7	Sikkim	1086	616	56.72
8	Tripura	5773	3663	63.45

(ii) Higher Secondary Stage

S. No	State	Total Teachers	Untrained Teachers	% of Untrained
1	Arunachal Pradesh	586	190	32.42
2	Assam	8540	5015	58.72
3	Manipur	1436	835	58.15
4	Meghalaya	609	308	50.57
5	Mizoram	528	332	62.88
6	Nagaland	224	165	73.66
7	Sikkim	428	232	54.21
8	Tripura	3130	1971	62.97

Source: NCERT 7th All India School Education Survey (Reference Date: 30th September, 2002)

a) Learning at school must be linked with learning outside the school

b) Need to learn goes beyond the book. The text book is only a resource and it should be

c) Combined with other resources

d) Learning must not be rote learning which masquerades scholarship

e) Making examination more flexible

f) Using education to inculcate participatory democratic identity

Requirements for implementing the 6 principles

a) Closer interaction between learner and the teacher

b) Creating a system where every subject has trained manpower

c) Reconceptualise education Reforms

d) Intensive Teacher Training Reforms

e) Administrative and Financial Reforms

f) Education to be made accessible to inaccessible domain

For improving the quality of education, specially during the formative years the child, the designated education institutes in every state need to set up teachers' training programme at all levels. Table 4 shows the large number of teachers who need to be trained and that can only be achieved with the help and collaboration of the institutes of higher learning in every state by providing their infrastructural facilities. The universities and the technical institutes can bring about tangible improvement in the teaching of Science and Mathematics and of the medium of expression of the language. The curriculum also needs to be rationalized. To bring about such reforms in a short time i.e. 2-3 years, a multi-pronged strategy in the following lines needs to be adopted:

1. Build a large resource group

2. Train higher secondary teachers as Resource Persons for teaching Science and Mathematics

3. Provide opportunity for interaction between the teacher and an established laboratory

4. Use Distance Mode of Education combined with contact and hand-on training sessions

5. Train teachers for communicative skills and abilities to interact with nursery and primary level children

D. *Skill-up gradation in Vocations relevant to the region and Integration of Vocational stream with Academic stream*: To address the problem of unemployment, students and youth need to be trained in skills and vocations that are relevant to the area and to their liking to enable them to earn a living when they leave the education system. For this, vocational stream has to be integrated with the general academic stream and the curriculum needs to be appropriately restructured and reoriented. Vocationalisation at the secondary level should provide entry both in the general stream as well as in the professional stream depending upon the student/learners' choice. Distance learning courses should also be available to enable the learners to learn at their own pace, time and place for the further up gradation of their skill and develop entrepreneurial capabilities.

E. *Extensive use of Information Technology in Education:* Extensive use of IT would enable sharing of areas of excellence and the know-how and expertise available in the various institute of learning in the region. Model lessons, especially in mathematics, science could be prepared centrally and beamed right up to primary school level. This could be the quickest way of overcoming the problems of difficult terrains, inaccessible localities and long distances. All centers of excellence should be interlinked.

The above reasons point us to reflect on our educational system. In order to bring about reforms, the fundamentals of the education system have to be remembered, realized and incorporated in the curricula. In order to bring about reforms, one needs to ask: Is there a difference between life and livelihood? Does life go beyond livelihood? What percentage of our life consists of livelihood? What happens to a person whose life and livelihood are the same? Pondering on these questions will provide us with certain clues on how to affect reforms.

References

Abraham, I. 2000. "International Trends, India-Seminar, 2000". 14 Dec., 2006. Cited in http://www.india-seminar.com/2000/494/494% 20itty % 20 abraham. htm

Altbach, P. G. 2005. Higher education in India *The Hindu*, 12 April 2005.

Cooper, C. 1991. Educational Guidance and Flexible Adult Learnings. *The Commmunication* 2(10).

Mansukham, S. 2004. Educational Changes and Continuity: Training and Development in Organizational Programme. *The Communication* 14(1).

Mehrotra, V.S. 2005. In: *Vocational Education and Training: Challenges and Strategies.* (Eds.) Sacheti, Verma & P.S.S.Mehrotra. Central Institute of Vocational Education, NCERT, Bhopal.

Menon, M.B. & V. Joshi. 1986. Designing the Instructional System for the Open University - A Perspective. *University News, Spl. Issue on Distance Education* 24(42).

Shaffila, Y. 1994. Accommodating Social Demand for Higher Education and the Cost-Effectiveness of Open Universities: A case Study. *Kakatiya. J. Distance Education.* 3(2).

Shah, A.M. 2002. Critical Success factors and Effective Strategies in the New Environment. *Vikalpa* 25(2).

Venugopal, V. & M. Srivastava. 2002. *Towards Virtualisation of Open and Distance Learning*. Kogan Page India Ltd., New Delhi.

3

Behavioural Studies on Birds and Conservation

LALITHA VIJAYAN

ABSTRACT

Studies on the behaviour of animals are of utmost importance in the conservation of species. Ecology of the communities and individual species has to be examined in terms of different activities to understand the requirements of the species. Foraging and breeding are the major activities while resting and flight maintenance are also essential, although the time spent might vary in different species or situations.

Bird community studies conducted along with my students in different habitats in the Western Ghats since 1995 showed how the species are associated with each other and the impact of habitat changes on them. Endemic and threatened birds of the Western Ghats were specific about selection of habitat and food in different seasons. Identification of such critical habitats is crucial for conservation and management.

The studies on the bird community and drongos at Thekkady, Periyar Tiger Reserve conducted by me during 1979 to 1982, examined the association of drongos with other species in the area, especially while foraging. They fed mostly in mixed foraging flocks which is beneficial to all the participants. The strategy adopted by them for ecological isolation to reduce competition was by selecting different strata and substrate of the vegetation, varying sizes of insects, and methods of feeding.

In the case of wetland birds also habitat selection is a major behaviour limiting their numbers. For the threatened birds such as the Siberian Crane and the Andaman Teal, habitat quality and food availability were critical. Habitat degradation with over abundance of grass and reduction in sedges resulted in spending long duration for feeding and less returns.

Breeding behaviour of birds is interesting because of their strategies such as when to nest and where to nest. Studies on the seasonality have shown the importance of climatic factors. Many of our land birds nest during summer. Nest-site is also crucial as selection of the site with suitable macro and micro-habitats determine the success of the brood. The habitat patches for the individual pairs or groups of breeding birds which are defended against intruders are termed as territories which play a vital role in the distribution and density of birds. Thus the strategies adopted by species are dependent on the environment and species specific constraints affecting the survival of the species.

INTRODUCTION

Behaviour and ecology are interrelated and behaviour of species is determined by the environment and, morphology and physiology of the species. Hence, studies on the behaviour of animals are of utmost importance in understanding the well being of the individual or species in a particular area. Birds have been extensively studied for documenting these phenomena and are considered as indicators of the environment (Canterbury *et al.*, 2000). The environmental or habitat quality can be assessed by documenting the habitat use (Martin & Karr 1986; Rosenzweig, 1991), especially for feeding as the time spent for feeding is negatively related mainly to the availability of food and hence is important in the conservation of species. Ecology of the communities and individual species has to be examined to understand the requirements of the species in terms of different activities. Foraging and breeding are the major activities while resting and flight maintenance are also essential, although the time spent might vary in different species or situations.

RESULTS & DISCUSSION

The studies on the bird community and drongos at Thekkady, Periyar Tiger Reserve conducted by me during 1979 to 1982, examined the association of drongos with other species in the area, especially while foraging. They fed mostly in mixed foraging flocks which is beneficial to all the participants (Vijayan, 1984). The strategy adopted by them for ecological isolation to reduce competition was by selecting different strata and substrate of the vegetation, varying sizes of prey (mostly insects) and methods of feeding (Vijayan, 1984). Strata and size of the insects selected corresponded with the size of the bird species. Competition was evident between two smaller species, the Grey Drongo, a local migrant and Bronzed Drongo, a resident species. When these two species fed together in mixed flocks, the Bonzed drongo came down from the top canopy giving way to the dominant Grey Drongo revealing the "compression hypothesis" (MacArthur & Wilson, 1967) or "Character displacement" (Cody, 1974) to reduce competition between these congeners.

Bird community studies conducted along with my students in different habitats in the Western Ghats since 1995 showed the community structure of birds in different forest types in the Nilgiri Biosphere Reserve (Gokula & Vijayan, 2000, 2007) and the impact of habitat changes on them (Vijayan *et al.*, 1998). Endemic and rare birds of the Western Ghats are specific to particular habitat; for instance, the Nilgiri Laughing Thrush and Nilgiri Pipit were adversely affected by the habitat alterations in Upper Nilgiris. Identification and documentation of such critical habitats is crucial for conservation and management (Vijayan & Gokula, 1999). Estimation of the extent of the habitat and population of the Nilgiri Laughing Thrush was done using the Remote sensing and GIS techniques (Vijayan *et al.*, 2001). Ecological studies on two other endemic birds, namely the Nilgiri Wood Pigeon and Nilgiri Pipit could identify the crucial factor in the survival of these species, namely the breeding habitat, *shola* forest in the former and wet shola grasslands in the latter (Umamaheshwari & Vijayan, 2003).

Bird communities of the Andaman Islands and habitat use of the endemics showed the importance of semi-evergreen and evergreen forests and their extent in determining the areas for biodiversity conservation (Vijayan *et al.*, 2005). The importance of these habitats was found true in the case of the rare and data deficient species, the Andaman Crane (Vijayan & Ezhilarasi, 2007). Bird communities along the elevation gradient in Teesta Valley, Sikkim showed high diversity of birds indicating the significance of mid elevation broad leaved hill forests and the need for protecting these areas for the conservation of biodiversity (Vijayan *et al.*, 2006a).

In the case of wetland birds also habitat selection is a major behaviour limiting their numbers, besides loss of habitat and persecution (Vijayan *et al.,* 2006b). For the threatened birds such as the Siberian Crane and the Andaman Teal, habitat quality and food availability were critical (Vijayan *et al.,* 2006c). Habitat degradation with over abundance of grass and reduction in sedges resulted in spending long periods for feeding but less returns in the Siberian Crane which would affect their return migration and successful breeding in the coming season (Vijayan & Vijayan, 2002). In the Andaman Teal the critical factor was the availability of the breeding habitat which was dependent on the water level and vegetation in the area (Vijayan *et al.,* 2006c).

Breeding behaviour of birds is interesting because of their strategies regarding when to nest and where to nest. Studies on the seasonality have shown the importance of climatic factors (Gwinner,1996); many of our land birds nest during summer (Vijayan *et al.,* 1998) while some of them do so in the rain shadow valley before the north-east monsoon (Nirmala & Vijayan, 2007). Nest-site is also crucial as site with suitable macro and micro-habitats determine the success of the brood (Gokula & Vijayan, 2001, 2003; Vijayan *et al.,* 1998; Ezhilarasi, 2007). The habitat patches for the individual pairs or groups of breeding birds are termed as territories (Nice, 1941) which play a vital role in the distribution and density of birds (Vijayan & Ezhilarasi, 2007). In the drongos, the number of breeding pairs or groups and the number of territories remained the same during the three years of study (1979-81) in spite of the addition to the population every year (Vijayan, 1984) and this might have allowed the presence of helpers at the nest in these species. Several environmental factors, mainly the availability of food and nesting sites determined the location and size of the territories. Thus the strategies adopted by species are dependent on the environment and species specific constraints ensuring the survival of the species (Newton, 1998).

References

Canterbury, G. E., T.E. Martin, D.R. Petit, L.J. Petit & D.F. Bradford. 2000. Bird communities and habitat as ecological indicators of forest condition in regional monitoring. *Conservation Biology* 14 (2): 544-558.

Cody, M.L. 1974. *Competition and the structure of bird communities. Monographs in population biology.No.7.* Princeton University Press, Princeton, New Jersey.

Ibid 1985. Introduction to habitat selection in birds. In : *Habitat selections in birds.* (Ed.) M.L. Cody. Pp. 4-6. Academic Press, London.

Gokula, V. & L. Vijayan. 2000. Foraging Pattern of Birds during the Breeding Season in thorn forest of Mudumalai Wildlife Sanctuary, Tamil Nadu, Southern India. *Tropical Ecology* 41(2): 195-208.

Ibid 2003. Foraging and Nesting Behaviour of Asian Paradise Flycatcher *Terpsiphone papradisi* in Mudumalai Wildlife Sanctuary, Tamil Nadu, India. *Forktail* 19: 142-143.

Ibid. 2007. Foraging strategies of birds in partitioning of food resources in dry deciduous forest of Mudumalai Wildlife Sanctuary, Tamil Nadu, India. *J. Sci. Trans. Environ. Technov.*1(1): 36-42.

Gwinner, E. 1996. Circannual clocks in avian reproduction and migration. *Ibis* 138: 47-63.

Macarthur, R. H. & E. O. Wilson. 1967. *The Theory of Island Biogeography*. Princeton University Press, Princeton, USA.

Martin, T. E. & Karr. 1986. Patch utilization by migrating birds: resource oriented? *Ornis Scand.* 17:165-174.

Morrison, M. L. 1986. Bird populations as indicators of environmental change. In: *Current Ornithology*. (Ed.) R.F. Johnston. Pp.429-451. Plenum Press, New York & London.

Nice, M. M. 1954. The role of territory in bird life. *Am. Midland Nat.* 26(3): 441-487.

Nirmala (Sr.), T. & L. Vijayan. 2007. Breeding birds of Anaikatty Hills. In: *Proc. National Conference on Biotechnology and Biodiversity*. February 23-25, 2007. Vellalar College for Women, Erode. Tamil Nadu. (Abstract).

Newton, I. 1998. *Population limitation in birds*. Academic Press, San Diego, USA.

Rosenzweig, M. L. 1991. Habitat selection and population interactions: the search for mechanisms. *Amer. Nat.* 137: S5-S25.

Umamaheswary, J. & L. Vijayan. 2003. Some Observations on the Behavioural Ecology of the Nilgiri Pipit *In: Proc. of 28th Conference of the Ethological Soc. of India,* Mundanthurai, Tirunelveli (Eds.) R. Annamalai, M, Narayanan & Juliet Vanitharani. Dept. Zool., Sara Tucker College, Tirunelveli, Tamil Nadu. pp 83-86.

Vijayan, L. 1984. *Comparative Biology of Drongos (Family Dicruridae, Class Aves) with special reference to ecological isolation*. Ph.D. Thesis. University of Bombay, Mumbai.

Vijayan, L. & N. Ezhilarasi. (2007). *Status and Ecology of the Andaman Crane. Final Report of the project.* Salim Ali Centre for Ornithology & Natural History, Coimbatore. Pp. 100+vii.

Vijayan, L. & V. Gokula. 1999. Impact of Human Interference on some rare Endemic Birds in the Upper Nilgiris, Tamil Nadu. Pp 127-134 In: *Endemic and Endangered Plant and Animal species of Eastern and Western Ghats.* (Eds.) G. Kumaravelu & K. K. Chaudhuri. *Proc. National seminar,* 22-23 March, 1999, Forest Department, Government of Tamil Nadu, Chennai.

Vijayan, L. & V.S. Vijayan. 2002. Ecology of the Siberian Crane in Keoladeo National Park, Bharatpur, Rajasthan during the winter of 1992-1993. In : *Birds of Wetlands and Grasslands. Proc. Salim Ali Centenary Seminar on the conservation of birds of the wetlands and grasslands.*1996 (Eds.) A. R. Rahmani & G. Ugra.

Vijayan, L., V. S. Vijayan. C. Venkatraman. & P. R. Arun. 1998. Breeding Strategies of Birds in a moist deciduous forest at Siruvani in the Nilgiri Biosphere Reserve. In: *Biosphere reserves and management in India.* (Eds.) R.K. Maikhuri, K. S. Rao & R. K. Rai. Pp. 179-185. G.B. Pant Institute of Himalayan Environment & Development, Almora, U. P.

Vijayan, L., V. Gokula. & S. N. Prasad. 2001. Study of the Nilgiri Laughing Thrush, Tamil Nadu, India. *OBC bulletin* 34: 18-19.

Vijayan, L., S. N. Prasad, M. A. R. Mamannan. & P. Kaushik. 2005. *Avifaunal Diversity of the Andaman Islands and their Conservation. (Final report of the project).* Salim Ali Centre for Ornithology & Natural History, Coimbatore. Pp.84.

Vijayan, L., S. Bhupathy, Bhoj Kumar Acharya, Joya Thapa & Basundhara Chettri. 2006a. An ecological study on mammals, birds, herpetofauna and butterflies in Teesta river basin, Sikkim. Carrying capacity study of Teesta basin in Sikkim Vol VII: Bilogical Environment- Faunal elements. CISMHE, Delhi University & Salim Ali Centre for Ornithology & Natural History, Coimbatore. Final report of the project.

Vijayan, L., V. S. Vijayan., S. N. Prasad. & S. Muralidharan. 2006b. Conservation of wetland birds in India. *J. Ornithol.* 147 (Suppl.) 24th International Ornithological Congress-Abstracts. Pp.80.

Vijayan, L., V. Murugan.V. & M. A Rajamannan. 2006c. Conservation of the Andaman Teal. *Threatened Waterfowl Specialist Group Newsletter* 15:55-59.

4

The Functional Organization of an Insect Society

RAGHAVENDRA GADAGKAR

ABSTRACT

Insect societies such as those of ants, bees, wasps and termites are characterized by reproductive division of labour and cooperative brood care. Thus these societies consist of one or a small number of reproductives (usually only queens but also kings in the case of termites) and a large number of sterile or nearly sterile workers. While the queens engage in laying eggs, all the tasks required for nest building, acquisition and processing of food and brood care are performed by the workers. The functional organization of such insect societies that leads to efficient division of labour is a topic of obvious interest. In this lecture I will describe experiments designed to understand the functional organization of one such insect society. I have chosen the primitively eusocial wasp *Ropalidia marginata* for these investigations. *Ropalidia marginata* is abundantly distributed in peninsular India and serves as an excellent model system. An important feature of this species is that queens and workers are morphologically identical and physiologically nearly so. How does an individual become a queen? How does the queen suppress worker reproduction? How does the queen regulate the non-reproductive activities of the workers? What is the function of aggression shown by different individuals? How and when is the queen's heir decided? I will show how such questions can be experimentally investigated and discuss what the answers to these questions tell us about the functional organization of this primitive insect society.

5

Status, Distribution, Food & Feeding of Malabar Spiny Dormouse (*Platacanthomys lasiurus* Blyth) in the Western Ghats of Kerala

JAYSON, E.A. AND K. M. JAYAHARI

ABSTRACT

Malabar Spiny Dormouse (*Platacanthomys lasiurus*), the endemic rodent found in the Western Ghats of South India was studied in the Peppara Wildlife Sanctuary, Thiruvananthapuram District, Kerala, to determine the status, distribution, food and feeding of the species. The species was recorded from ten protected areas in Kerala and from twenty one forest ranges. Trapping studies indicated extremely low densities in the protected areas of Kerala. It feeds on a variety of food items available in the habitat. Common food plants of the species were *Terminalia bellerica, Persia macrantha, Hydnocarpus pentandra, Tamrindus indica, Bombax ceiba* and *Shumanianthus virgatus*. Favorite food items included endemic genus like Piper, exotic species like *Theobroma cacoa* and *Anacardium occidentale*. Foraging decisions were influenced by the predator pressure. Sexual dimorphism existed in the anti predator behaviour and females were bolder than males. Two major types of foraging movements were observed in the species. One type is limited to a distance of 30 m from the nest and other beyond this distance. In the case of males, short feeding bouts were observed during the short distance foraging and long feeding bouts in long distance foraging. In females, considerably long feeding bouts were recorded during the long and short distance foraging movements. The male moved to a maximum of 1 km during foraging and the female 80 m and the feeding range of the species was about one km. Males carried out the nest hole maintenance activity by gnawing the nest entrance for removing the continuous growth of bark. Female was never observed carrying out the nest maintenance activity. The behaviour of nest hole maintenance can be attributed to the antipredator strategy. *Lagerstroemia microcarpa* is a preferred nesting tree of Malabar Spiny Dormouse. Birds of prey and owls are the main predators recorded. The Malabar Spiny Dormouse is highly selective in choosing the microhabitat and high degree of preference is given to the dense canopy cover and crown density. Preference for canopy cover is also an antipredator strategy.

6

The Science and Art of Monitoring Wild Tigers

ULLAS KARANTH, K.

ABSTRACT

Science essentially involves confronting our 'models' of nature, against data from the real world, in a continuous effort to refine our knowledge of the system under study. Scientific monitoring of animal populations is therefore best viewed in the context of questions such as why monitor, what to monitor and how to monitor. This talk will cover reasons for which wild tiger populations need to be monitored, what are the 'state variables' that we need to monitor and how we can go about monitoring these parameters of interest. Measurement of parameters such as habitat occupancy and population abundance as well as vital rates that drive changes in these variables are examined in the context of modern survey sampling methods, parameter estimation and modeling approaches. It will be shown that measuring these ecological and demographic parameters requires a sound understanding of tiger behavior also.

Next the critical issue of integrating knowledge of tiger ecology with the local social context and the availability of human, technical and financial resources for population monitoring is discussed. Thus, we see that inevitably, the science of monitoring tigers turns into something of an art form too. The challenges involved in the social process of integrating scientific monitoring data on tigers into practical conservation or management measures in the real world, is illustrated based on examples from long term work in India. A vision of how ecological knowledge about tigers can be applied to managing wild tiger populations across large landscapes is presented in the context of the new WCS initiative, *Tigers Forever*.

7

Social Cognition in Primates From Mental Representation to Distributed Cognition

ANINDYA SINHA

ABSTRACT

Social primates appear to be knowledgeable about one another's behaviour to different extents. But do they know as much about one another's beliefs and intentions? Are they adept at recognizing the similarities and differences between their own and others' states of mind? Attribution of mental states to other individuals could manifest itself in diverse situations as, for example, when individual animals closely observe the actions of others, when they interact competitively, or when they deceive each other in the social sphere. The alternative perspective of primate distributed cognition, however, disregards the ability of an individual to have observationally invisible mental states and only recognizes communicative interactions and the behavioural dynamics within the entire group as manifestation of the socio-cognitive complexity that individuals are alone capable of displaying. This talk will examine some of these theoretical and philosophical issues in primate cognitive psychology, ranging from mental representation to distributed cognition. I will also use these perspectives in attempting to understand the cognitive mechanisms potentially underlying complex phenomena such as tactical deception, acquisition of social knowledge, phenotypic flexibility in social behavioural strategies and the appearance of behavioural traditions in wild bonnet macaques, a primate species found commonly in peninsular India.

8

Signallers and Receivers Sensory Ecology of Acoustic Communication in Crickets

ROHINI BALAKRISHNAN

ABSTRACT

Male crickets use acoustic signals for long-distance mate attraction. Each cricket species has a unique calling song, used by females to recognize and localize conspecific males. Field cricket songs have wavelengths of several centimeters, which pose a difficult problem for these small animals in terms of localizing males, particularly in choruses. The chorus structure in terms of spacing, call intensity and timing, are important determinants both of the ability of males to synchronise their calls and of female mate-finding strategies.

Male crickets often call within hearing distance of each other in the field. This results in chorusing patterns such as synchrony, where two or more males produce chirps simultaneously. Acoustic synchrony between calling males has been speculated to evolve as a result of in-built neural biases that cause females to prefer males whose chirps synchronise with their neighbours but lead them by a small amount of time (the precedence effect). This hypothesis of the evolution of chorusing depends crucially on two factors: the characteristics of the neural circuitry (oscillator) underlying song production and the presence of selective attention (restriction of interaction to a subset of calling neighbours). The bush cricket, *Mecopoda* spp. (Chirper) exhibits acoustic synchrony, achieved by chirp-by-chirp resetting of the song oscillator and by changes in chirp rate. Our examination of chorus structure in the field shows that males typically interact with one or two calling neighbours.

How female crickets locate one male in a chorus is still unexplained. To this end, we have characterized field cricket choruses in terms of male spacing, movement and calling. Using closed-loop phonotactic experiments on freely walking crickets in the laboratory, we have explored the rules of sensory and motor behaviour and used these in a simulation of walking phonotaxis.

9

Recent Trends in Ethology

JULIET VANITHARANI

ABSTRACT

Ethology, the biology of behaviour has been a fascinating area of biological investigations during past few decades. In recent past the information gathered by reputed ethologists on social behaviour of insects, fishes, birds and human beings are very helpful in the application of these principles for the welfare of the humanity as well as for the environmental sustainability. Ethology differs from the study of Animal Behavior, in that animal behaviorists generally are interested in learned behaviors while ethologists concentrate on innate behaviors.

Ethology has developed as a major field of research in biology since the 1970s. It is interrelated with various disciplines such as ecology, physiology, endocrinology, neurobiology, genetics and evolution. The planet earth is enriched by its biodiversity and has become wonderful due to the fascinating pattern of behaviour of the diverse organisms. Recent advancement in ethology has revealed a variety of interactions between plant and animals. Most animals show direct dependence on the plants. There are also plants that are directly dependent on animals for their continued existence in their natural habitats.

Much of the work in ethology revolves around problems in animal communication. Pheromones are chemical "perfumes" that one animal emits to influence the behavior of another. Chemical communication has been found in nearly all organisms investigated, including fungi, bacteria, protists, and plants. Among the animals, nearly all use pheromones at one level or another, with the notable exception of most birds. Ethologists in recent years are unraveling the neuro-endocrine basis of behavior to explain the proximate reasons behind why animals act the way they do. The topics like communication, social behaviour, reproductive behaviour, plant animal interaction, neuro ecology, biological rhythm, and behavioural genetics are the major fields at present which ultimately help in biodiversity conservation and the maintenance of sustainable ecosystem. Very complex behavior patterns can be passed on through the genes 'The future of biology is in molecular biology and the future of molecular biology is physiology and behaviour. It is only with research on ethology that India can ever hope to be in the cutting edge of world class research in biology.

❑❑❑

10

Parsimony in Insects

GEETHA BALI

ABSTRACT

Insects are considered the most successful in the animal kingdom. Many factors contribute to their success. One major factor that operates strikingly is the economy and optimal use of energy. One can give several examples taking the life of a single insect be it locomotion, respiration, excretion, organization of the nervous system, etc. This phenomenon of optimal use of energy is clearly evident in reproduction and reproductive behaviour of insects. Insects have evolved several mechanisms to operate in a fool proof way but with great success and highest efficiency with reference to utilization of energy. This trend can be seen not only in individual insects but at the population level also. Contributing towards this are a battery of sensory structures which are extremely sensitive and can perceive a variety of signals very accurately from the environment. The second noteworthy feature is the redundancy and multiple regulatory mechanisms so that one takes over if the other one fails and also makes fine tuning and precise modulation possible. The third factor is the employment of a variety of chemical regulatory molecules orchestrating behaviour in a most complicated way and regulating metabolic activities in a complementary way to support the needs in performing timed, sequential and complex behavioural acts. These along with the anatomical organization of the nervous system in insects present the most interesting arena nature has evolved.

11

Localisation of Prey by Young and Adult Indian False Vampire Bats

RAGHURAM, H AND G. MARIMUTHU

ABSTRACT

The Indian false vampire bat, *Megaderma lyra* detects and captures prey (e.g. frogs) mainly by listening to the rustling sounds that are created when the prey is on ground. In addition, echolocation and vision are also involved. Mothers of *M. lyra* take meticulous care of their young. At the age of about seven weeks, juveniles exhibit gleaning flights. Among eight mother-young pairs, four young (group A) began to capture frogs at the age of 62 days, although their mothers continued to suckle until they were 85 days old. The mothers of the remaining four young (group B) stopped suckling, although their 60 days old young did not exhibit prey capture by themselves. However, mothers transferred their partly consumed prey, only when their young (B) solicited. Based upon the size of frogs, the transferred morsels varied from entire frogs to only single hind limbs. Such selective food-transfers may be due to the limitations on the young to handle the prey parts. The mothers stopped food-transfers, in spite of their young's solicitation at the age of 67 days. When mothers repeatedly avoided the approaches of young for food, the latter exhibited kleptoparasitism (stealing food) from their mothers until the age of 75 days. These four young (B) began to capture frogs at 70-81 days old. Food-transfer and kleptoparasitism did not occur with the young of group A.

The non-food-transfer mothers always captured prey prior to their young. This may be a method of tutoring, facilitating observational learning by the young (A) whereas, food-transfer mothers always captured prey, only after their young (B) completed their captures. This may be to avoid kleptoparasitism.

SECTION - 2
BEHAVIOURAL ECOLOGY

12

Molecular Approaches in Behavioural Ecology

SAIKAT CHAKRABORTY

INTRODUCTION

Molecular tools generally refer to molecular genetic tools. Although the first use of these tools began more than four decades ago, when Lewontin & Hubby (1966) used allozymes to find out the amount of genetic variation present in wild populations of *Drosophila*, the lack of expertise and high cost of these tools have prevented their widespread use among ecologists in general and behavioural ecologists in particular. But recently the accessibility and costs of these techniques have come down a lot and with the right kind of collaboration, lack of expertise is no longer an issue.

As was mentioned above, the Lewontin & Hubby's (1966) investigation was done using allozymes. Although still an option, user-friendly neutral nucleotide markers are more preferred by investigators nowadays. The discovery of the polymerase chain reaction (PCR) in 1983 by Kary Mullis was the single landmark in this field, if one has to mention one. The use of molecular markers has totally revolutionized the field of ecology. It has provided ecologists the tool to ask questions, which were hitherto impossible to tackle and also to answer questions with much more precision and accuracy than was possible till now. It has toppled long held beliefs, sometimes, romantic ones, by ecologists.

With collaboration becoming more of an option and with the techniques becoming cheaper, more and more investigators are likely to use molecular tools to address their questions. In fact now there are well-formed commercial enterprises, which will do everything starting from isolation of the loci, creating primers and genotyping the samples. Even they will write the "primer notes" for your target journal.

The analysis of the data obtained in this way is even easier as long as one is familiar with the theory. There are batteries of user-friendly, industry standard free softwares, one or more of which can be used to analyze the data. With computers becoming more and more powerful, these softwares can do more "accurate" simulation based parameter estimates using Bayesian, Maximum Likelihood and Markov Chain Monte-Carlo methods, rather than approximations using summary based parameter estimates found in earlier literature.

In summary, there hasn't been a better time to do ecological sciences. Now let us have a look at the tools we have and the kind of questions that can be answered by using them.

AN OUTLINE OF MOLECULAR TOOLS

Generally molecular tools tend to tap the genetic variation present in individuals on small scale and the population at large scale. Both nuclear as well as organellar (mitochondrial and plastid) DNA can be used and one is preferred over the other depending on the question being asked. Molecular tools can be classified as either dominant or co-dominant markers, depending on whether both the alleles at a locus come as the data or the presence of the "dominant" allele masks the "recessive" one.

Dominant Markers typically generate data from multiple loci at a time. In this case, no prior knowledge of the loci is required as random primers are used and the development time is also short as a result. The random markers generally amplify multiple regions of the genome, as a result they show considerably high polymorphism. As was mentioned earlier, the main drawback of this kind of marker is that only one allele at a locus can be identified. This makes calculation of allele frequencies particularly difficult, unless Hardy-Weinberg equilibrium for the locus is assumed. Below is a brief outline of some of the dominant markers.

(i) *Random Amplified Polymorphic DNA (RAPD):* RAPD involves the use of random primers usually 10 base pairs (bp) in length. One would expect a random10 base length sequence to repeat once every 1048576 (4^{10}) bases. In each PCR (Polymerase Chain Reaction) reaction, one kind of primer is used. The primer anneals with the complementary regions of the two DNA strand and the intervening region is amplified. If more than one such region is present in the genome then more than one band will arise once gel electrophoresis is performed with the PCR product.

(ii) *Amplified Fragment Length Polymorphism (AFLP):* AFLP involves first digestion of the DNA with two different restriction enzymes (details about restriction enzymes is given in the section dealing with RFLP). The overhanging ends created in this way are then annealed with pre-synthesized "linker" DNA with sequences complimentary with the overhanging region. Primers are designed that are complimentary to this linker region and PCR is done. The resultant product is then run in gel to separate them according to size.

Co-dominant Markers are always preferred over dominant markers as knowledge about both the alleles in a locus for a diploid individual can be gained. This is essential as was mentioned earlier, for the calculation of allele frequencies, which is the first step for addressing most molecular ecological questions. Following is a brief outline of some of the co-dominant markers:

(i) *Allozymes*: This is the oldest marker to be found in the literature dealing with molecular ecology. Allozymes refer to various forms of an enzyme with the same function but different amino acid sequences. According to the neutral theory of molecular evolution most of the amino acid replacements in a protein won't affect it's functionality to the extent that the survival and reproduction of an individual will be jeopardized. In fact, this model says that most mutations at a locus are neutral from the perspective of fitness. Allozymes are co-dominant in the sense that if one individual has more than one form of the enzyme i.e. with different amino acid sequences the differences will show up when electrophoresis is done and the enzyme is visualized using the right kind of substrate and dye.

Allozymes provide with a conservative estimate of the available genetic variation as only non-synonymous mutations turn up as data. Using DNA data instead of proteins definitely gives much more resolution of the available genetic variation.

(ii) *Restriction Fragment Length Polymorphism (RFLP):* The DNA is digested with one or more restriction enzymes and the fragments obtained are run on gel. Restriction endonuclease enzymes are a class of enzymes obtained from bacteria e.g. EcoRI obtained from *Escherichia coli* and BamHI from *Bacillus amyloliquefaciens*. They cut the DNA where they encounter a specific sequence. In cases of most restriction enzymes this region is palindromic, e.g. EcoRI cuts at G|AATTC (the complementary site being CTTAA|G; | indicates the position where the DNA is cut). So any new inclusion or exclusion of restriction sites will show up as genetic variation.

(iii) *Single Nucleotide polymorphisms (SNPs):* As the name suggests SNPs refer to any of the alternative bases that can be present in a single nucleotide position i.e. whether A, T, C or G. Generally a single SNP locus is typified by two alternative bases i.e. alleles. SNPs are slowly becoming more popular.

(iv) *Minisatellites*: These are 10 to 100 bp tandem repeats present in the genome.

(v) Microsatellites are also known as STRs (Short Tandem Repeats) or VNTRs (Variable Number of Tandem Repeats) or SSRs (Simple Sequence Repeats). These are 1 to 6 bp tandem repeats and is by far the most widely used and recommended molecular marker. An example of a microsatellite locus with dinucleotide repeats will be ACACACACACACACACAC (adenine and cytosine repeated 9 times, also written as $(AC)^n$). Microsatellites are typically characterized by conserved 5′ and 3′ flanking regions, which helps in designing primers to perform PCR. In fact, using primers developed for one species, loci in different species have been amplified.

(vi) *DNA sequences:* By far the molecular marker with the highest resolution but extremely expensive at the same time is DNA sequences. The technique involved in sequencing DNA was developed independently by Frederick Sanger in England and Maxum and Gilbert in USA in the early 70s. This technique has been fully automated. Although not feasible for population level studies, systematists and phylogeneticists routinely use this technique.

MOLECULAR TOOLS IN ADDRESSING QUESTIONS IN BEHAVIOURAL ECOLOGY

(i) *Parentage Analysis:* A wide body of investigation can be grouped under parentage analysis. Parentage analysis is defined as the process of assigning offsprings to their respective parents, both father and mother. This has helped scientists to estimate the exact fitness for each individual, understanding the mating types e.g. monogamy, polygamy, polyandry, promiscuity etc. All of the above mentioned markers can be used for parentage analysis but microsatellites are the most recommended owing to their high variability and accuracy in scoring. SNPs can also be used for this purpose, although the number of loci to be taken into consideration, to obtain the same amount of resolution, is on the range of several orders of magnitude more compared to microsatellite loci (Jone & Andren, 2003).

Undoubtedly, the most astonishing revelation that molecular studies have brought us is the widespread prevalence of extra pair fertilization (APF) among birds. About 90% of the avian species are socially monogamous. Parentage studies using molecular markers have shown the presence of extra pair brood in 90% of the species studied (Griffith *et al.,* 2002). On an average over 11% of the offspring on a nest arise from extra-pair paternity. This means that although socially monogamous, they are in fact genetically polygamous (Pike & Petrie, 2003).

(ii) *Relatedness Studies:* Molecular markers have been used o measure the amount of relatedness among individuals in different groups across taxa. This is essential for testing kin selection theory. Using this technique kin selection has been confirmed as the reason behind the presence of helpers in the nest of several birds. At the same time, based on relatedness studies, kin selection has been withdrawn as the sole cause for the evolution of eusociality among social hymenoptera, although the theory was initially developed to explain this phenomenon. Detailed microsatellite analysis of the sperms stored in female hymenopteran's spermathecae has shown that the females mate with multiple males, extending up to or even more than 20 males in some species, drastically bringing down the average relatedness in a colony.

(iii) *Foraging Ecology:* Molecular markers have thrown new light on the habits of illusive species like Coyote (*Canes latrans*) and are likely to provide with new understandings about the foraging ecology of little known Indian species like stripped hyena (*Hyaena hyaena*) and so on. DNA from severely mutilated carcasses and tough parts of the prey species/plants present in scats/dung can be used to identify the range of species taken by a predator/herbivore as food.

(iv) *Sex-biased Dispersal:* Using either thorough knowledge about sex-specific markers or rigorous population genetic techniques, both analytical and simulation based, the difference between the ability to disperse among the sexes in a species can be studied.

CONCLUSION

The above can be treated only as a very brief outline of the kind of questions in behavioural ecology that can be addressed using molecular genetic tools. The reference includes some good reviews that deal with the above-mentioned topics in further details. In summary the cost of using these tools are bound to come down in future making them more accessible. As a result behavioural ecology as a field is going to get revolutionized with this new multidisciplinary collaborative approach and will make us more aware about the extraordinary behaviours of our lesser-known friends.

References

Griffith, S. C., I. P. F. Owens & K. A. Thurman. 2002. Extra pair paternity in birds: a review of interspecific variation and adaptive function. *Molecular Ecology* 11: 2195-2212

Jones, A. G. & W. R. Andren. 2003. Methods of parentage analysis in natural populations. *Molecular Ecology* 12: 2511-2523

Lewontin, R. C. & J. Hubby. 1966. A molecular approach to the study of genetic heterozygosity in natural populations. II. Amounts of variation and degree of heterozygosity in natural populations of *Drosophila pseudoobscura*. *Genetics* 54: 595-609.

Pike, T. W. & M. Petrie. 2003. Potential mechanisms of avian sex manipulation. *Biological Reviews* 78: 553-574.

Suggested Readings

Avise, J. C., A. G., Jones, D. Walker, D. & J. A. De Woody 2002. Genetic mating systems and reproductive natural histories of fishes: Lessons for ecology and evolution. *Annual Review of Genetics* 36: 19-45

Blouin, M. 2003. DNA-based methods for pedigree reconstruction and kinship analysis in natural populations. *Trends in Ecology and Evolution* 18: 503-511

Fitzpatrick, M. J., Y. Ben Shahar, H. M. Smid, L. E. M. Vet, G. E. Robinson & M. B. Sokolowski. 2005. Candidate genes for behavioural ecology. *Trends in Ecology and Evolution* 20: 96-104

Freeland, J. R. 2005. *Molecular Ecology*. John Wiley and Sons Ltd., London.

Neff, B. D. & T. E. Pitcher. 2005. Genetic quality and sexual selection: an integrated framework for good genes and compatible genes. *Molecular Ecology* 14:19-38

[illegible]

[illegible]

[illegible]

[illegible]

Hiscock, [illegible]

13

Habitat Use and Response of Brown-hawk Owl, *Ninox scutulata* to the Broadcast of Conspecific Calls

BABU, S. AND E. A. JAYSON

ABSTRACT

Tropical wet evergreen and tropical moist deciduous forests were the preferred habitats of the Brown-hawk Owl. Detection rate and response to the broadcasted calls were highest during the breeding season and in all the census points, more than one individual *i.e.* either in pair or three individuals were detected. Besides conspecific species, some heterospecific species were also found to respond to the broadcasted calls of Brown-hawk Owl, where the Brown-hawk Owl was not detected during the Initial Quite Listening survey. Moreover, approach of the sympatric species was observed only in the riparian strip and in rest of the habitats they responded only with calls. No significant seasonal difference was observed in the habitat utilization but seasonality was observed in the response and the approaching distance to the conspecific calls broadcasted. Additionally, the response to the conspecfic calls indicated that the owls were responding to the conspecfic calls throughout the year.

INTRODUCTION

The vocalization of several species of birds are communication signals that convey information about the bird's quality and fitness, which is vital in maintaining the status of the territory and also to elicit mate attraction, which will lead to production of successful progenies (Galeotti & Pavan, 1993; Appleby & Redpath, 1997; Galeotti, 1998). This is highly obvious in some non-oscine species like owls. Recognition of neighbours and strangers also exists in owls. The defensive mechanism in owls towards the intruders differs between the sexes (Galeotti *et al.*, 1997; Appleby *et al.*, 1999). Considering these behavioural traits, several studies have used the recorded conspecific calls of owls for eliciting response from local owls for census. Later it has been adopted as an effective census tool for the group of elusive predatory birds (Forsman *et al.*, 1984; Kavanagh & Peake, 1993; Debus, 1995; Zuberogotia & Campos, 1998). Even though, in some studies owl detectability showed difference, the broadcasted conspecific call was not an effective sampling method for censusing the floaters (young, dispersing individuals or vagrant adult that search for a territory to initiate the life) and paired territorial owners together (Rohner, 1997).

Brown hawk Owl is one of the thirty-three species of owls found in India and the only representative of the genus *Ninox*. The distribution of this species is all over the oriental region, which includes India, China, and Bhutan. There are five sub species known to occur in the Indian subcontinent. We studied the responsive behaviour of the sub species *Ninox scutulata hirsuta*. It was selected for the study, because of the immediate response to its conspecific calls so that the data on interspecific interaction can be estimated thoroughly to make strong interpretation on repertoire and evolution of such behaviour. Behavioural and ecological aspects of forest owls in the Indian subcontinent are scanty (Neelakandan, 1980; Rasmussen & Collar, 1998; Ramanujam, 2004; Ishtiaq & Rahmani, 2005). No studies on owls have reported the level of interaction with its sympatric organisms that have overlapping geographical ranges. Olson *et al.*, (2002) & Debus (1996) extensively studied the vocal behaviour with specialization on dueting in Southern Boobook Owl (genus *Ninox*). Although studies have been carried out on the vocal description and behaviour of the genus *Ninox* in the temperate countries, population status and distribution of the Indian species, *Ninox scutulata* are unknown. In this paper we are reporting the (1) efficiency of the survey methods (2) distribution and population status of the Brown hawk Owl and (3) the inter and intra specific behaviour and habitat utilization.

MATERIAL & METHODS

Study area: We visited all the protected and reserved forests of Tamil Nadu and Kerala States (Fig 1) for the survey of owls in the southern Western Ghats. Data from five protected areas and two reserve forests in Kerala and four protected areas and one reserve forest in Tamil Nadu were choosen for the study and analysis (Table 1).

Table 1. Locations surveyed for the study in Kerala and Tamil Nadu

S. no	Protected areas	Sq. Km	Number of census points surveyed	Major vegetation types in the study locations
	Kerala			
1.	Peechi-Vazhani Wildlife Sanctuary	126	16	Moist deciduous forest, Wet evergreen and Teak plantation
2.	Nelliampathy Reserved Forest	205	6	Wet evergreen, Coffee and tea plantation
3.	Parambikulam Wildlife Sanctuary	274	4	Teak, Wet evergreen and Moist deciduous forest
4	Thattekkad Bird Sanctuary	25	4	Riparian, teak and Moist deciduous forest
5	Chimmony Wildlife Sanctuary	75	6	Moist deciduous and semi evergreen
6	Chinnar Wildlife Sanctuary	90	4	Dry thorn forest and riparian forest
7	Vazhachal Reserved Forest	413	6	Wet evergreen, moist deciduous and teak plantation
	Tamil Nadu			
1	Palni Hills	435	6	Sholas, plantations and low altitude dry forests
2.	Indira Gandhi Wildlife Sanctuary	958.60	12	Wet evergreen forest, teak plantation, dry deciduous forest and low altitude dry forests

contd....

table contd....

3.	Srivilliputtur Grizzled Squirrel Sanctuary	487	6	Dry deciduous forest, wet evergreen and plantations
4.	Kalakkad Mundanthurai Tiger Reserve	895	5	Wet evergreen, dry deciduous forest and riparian

Survey design : Even though several factors both spatially and locally influence the distribution and habitat selection of owls in the tropical forests, we selected the vegetation type as the main criterion for the selection of census points across different protected areas. Stratified random sampling method was adopted to estimate the number of sampling points needed in each habitat type. Each protected area surveyed was spatially gridded using MAP INFO software version 7 at 3.1 km interval. Number of census points per vegetation type was proportionately allotted to get maximum sampling in the dominant vegetation. Surveys were conducted in 75 locations in the southern Western Ghats, using Dusk watch, Initial Quite Listening, Broadcasted call and Spotlighting as the main methods

Dusk-watch: A period of 15 minutes was selected as an ideal time period (after many trials in the field) for recording the presence of owls in a location. Fifteen minutes was spent at selected sites at dusk, watching and listening quietly for the calls or movements of the owls and other animals in the surrounding location. A 55-watt, hand-held spotlight was used occasionally to check the identity of animals, but this was done sparingly to avoid disturbance. All the animals seen or heard were recorded.

Spontaneous calls: A standard period of 15 minutes was spent on arrival at each site, watching and listening quietly for calls or movements of owls and other animals in the surrounding forest. This is an efficient method to record the presence or absence of Brown hawk Owl in census points before call play back sequences which helped in deriving conclusions on interspecific as well as intra specific behaviour.

Broadcasted tape call: Call play back method is an effective and informative sampling method for both diurnal as well as nocturnal birds. The calls of conspecific owls were collected and a standard call playback sequence was constructed. This was of 2 minutes or 1 minute listening period. These sequences were broadcasted at selected sites using 10-watt speakers, at a volume of about 120% of the owls' natural level. While recorder was playing the observer remained several meters away to avoid distraction from tape noise. All owls and other wildlife seen or heard were recorded, along with basic information on wind, temperature and other weather conditions.

Spotlighting: After the completion of the call playback method, 10 minute spotlighting session was conducted to check for any owls, which may have flown in silently and to survey arboreal mammals or other nocturnal wildlife. During this session, the observer walked for about 100 m in each direction along the track, searching adjacent forest by listening and using a hand-held spotlight. All animals seen or heard were recorded. No surveys were carried out on days with heavy wind and rain to avoid bias in sampling.

BEHAVIOURAL OBSERVATIONS

Latency: Time taken to elicit vocal or physical response from the initiation of broadcasting, was measured in seconds.

Approaching distance and height: Distance from the centre of the census point and owl perched tree was measured in meters using marked ropes and height of the owl presence.

Type of detection: Recorded whether the first detection of owl was visual or acoustic.

Mode of response: Either physical or vocal or attacking the observer

*Habitat Description:*Point Centered Quadrat (PCQ) method was adopted to sample the vegetation structure and tree diversity in response sites (Mueller-Dombois & Ellenberg, 1974). PCQ is one of the commonly used methods for estimating the bird habitat, in which four distances were measured at each sampling point. In each quarter thus created, the distance between the nearest tree and the sampling point was measured. The distance of the four trees in each quarter was averaged and the density was calculated.

RESULTS

Status and distribution: This elusive species has been reported from most of the sanctuaries and reserved forests of Kerala and Tamil Nadu except Chinnar Wildlife Sanctuary. From the low altitudes sighting was negative (Table 2). Mid altitude (300 –1200 msl) forests, which included tropical semi evergreen, tropical moist deciduous forest and wet evergreen forest hold good population of Brown hawk Owl (Table 3). Out of 75 sites, the presence of Brown hawk Owl was recorded in 19 (26.03%) sampling points, which indicates the rarity and scattered distribution of the Brown hawk Owl.

Table 2. Distribution and detection rate of Brown hawk owl in different protected areas of Southern Western Ghats

S.No	Protected areas	Number of census points (a)	Points where owls were detected (b)	% of detection in different protected areas C= b/a X 100
	Kerala part of Western Ghats			
1.	Peechi-Vazhani WLS	16	6	37.50
2.	Chimmony WLS	6	1	16.66
3.	Thattekkad bird sanctuary	4	2	50.00
4.	Parambikulam WLS	4	1	25.00
5.	Nelliampathy Reserved Forest	6	4	66.67
6.	Vazhachal Reserved Forest	6	4	66.67
7.	Chinnar WLS	4	0	00.00
	Tamil Nadu part of Western Ghats			
8	Palni hills	6	0	00.00
9	Srivilliputtur WLS	6	1	16.67
10.	Indira Gandhi WLS	12	1	08.33
11.	Kalakkad-Mundanthurai Tiger Reserve	5	1	20.00
	Total	75	21	

The distribution pattern of Brown hawk Owl showed that the Kerala part of the Western Ghats supports a viable population in most of the protected areas (Table 3). Owls were detected in most of the census points of the two reserved forests such as Vazhachal (66.67%) and Nelliampathy (66.67%). Most of the climatic factors such as temperature, rainfall and wind velocity were found to influence the detection probability of Brown hawk Owl. Among them rainfall and wind velocity were the major factors, which obstructed efficient sampling.

Table 3: Detection rate of Brown hawk Owl in different vegetation types of southern Western Ghats

S.No	Vegetation types	Number of census points (a)	Points where Owls were detected (b)	% of detection in different vegetation types ©C= b/a X 100
1.	Moist deciduous forest	32	11	34.37
2.	Semi evergreen forest	3	2	66.66
3.	Wet evergreen forest	11	4	36.36
4.	Man-made plantations	15	3	20.00
5.	Dry deciduous forest	11	1	09.09
6.	Dry thorn forest	3	0	00.00
	Total	75	21	

Efficiency of survey methods: Out of the four methods applied to locate the owls in the sampling points, spontaneous call and broadcasted tape call yielded maximum detection while ignoring other climatic parameters. A total of 143 call sequences of Brown hawk Owl were broadcast across seventy-five census points and response was obtained for 41 call sequences. Out of which, during the 35 call sequences, the presence of owl was confirmed. Among these 35 call sequences only 75% of them got a response and in nine sequences the response was not received. Moreover, the broadcast call yielded 15 new records where the presence of this elusive species was not known earlier.

Interspecific and intra specific responses: The Collard's scops Owl and the Barred jungle Owlet responded more to the broadcasted calls of Brown hawk Owl. Among these two, Collard's scops Owl responded for 16 call sequences but Brown hawk Owl did not respond to any of the Collard's scops Owl calls. Similarly Brown hawk Owl responded rarely to the Barred jungle Owlet call at a considerable distance (100-

150 m) from the broadcasting station. Barred jungle Owlet also responded to the Brown hawk Owl calls during 11 call sequences (Table 4).

Immediately after its conspecific call, the Brown hawk Owl responded within a short period of interval (5-10 seconds). But at certain occasions they responded one minute after the initiation of call sequences. This showed the variations of individuals in recognition of conspecific intruders. Similarly on few occasions, the Brown hawk Owl responded while broadcasting in the census station (7.32%). Even though the presence of Brown hawk owl was known during the 35 call sequence, it responded only to 26 call sequences (75%) and in nine sequences Barred jungle Owlet and Collard's scops Owl responded. Only in one site no response of any of the species was observed. More than 60% of the response of Brown Hawk Owl was within 20 seconds and Barred Jungle Owlet took 40 seconds for responses. Latency of calls recorded is shown in Tables 4 & 5.

Table 4. Latency of Barred jungle owlet's response to calls of its congeneric species

S.No	Seconds	Latency to Barred Jungle Owlet calls	Relative value of interspecific calls
1.	While broadcasting	0	00.00
2.	5-10 seconds	2	18.18
3.	11-20 seconds	2	18.18
4.	21-40 seconds	4	36.36
5.	41-60 seconds	3	27.27
6.	60-120 seconds	0	00.00
	Total	11	100.00

Table 5. Latency of Brown hawk Owl's response to its conspecific call

S.No	Seconds	Latency to conspecific calls	Relative value of Interspecific calls
1.	While broadcasting	3	7.32
2.	5-10 seconds	16	39.02
3.	11-20 seconds	10	24.39
4.	21-40 seconds	6	14.63
5.	41-60 seconds	3	7.32
6.	60-120 seconds	3	7.32
	Total	41	100.00

Approaching distance and height: Approaching distances of Brown hawk Owl between the breeding and non-breeding seasons showed a significant difference (Fig. 1). Similarly the height occupancy while defending the intruder of conspecific call differed between the breeding and non-breeding seasons. The reasons for such behaviour may be the territory ownerships and maintenance of breeding pair. They twice attacked the observer when the approaching distance was near to the broadcasting station *i.e.* 0-5 m distance at a low height (4-8 m). They were very aggressive during the breeding season than in the non-breeding season.

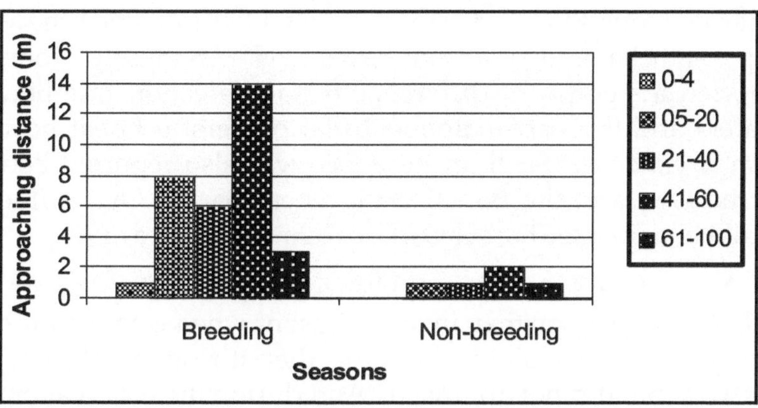

Fig. 1. Approaching distances of Brown hawk Owl during the breeding and non breeding seasons.

DISCUSSION

Application of the call playback method using the conspecific calls has proved to be an effective sampling method for many of the owl species (Kavanagh & Peake, 1993; Debus, 1995; Zuberogoitia & Campos, 1998; Martinez & Zuberogoitia, 2002) but in some of the species it is ineffective (Martinez & Zuberogoitia, 2002). However, our study is supporting the call playback method as an effective census method in the tropical forests, which enhanced the detection rate of Brown hawk Owl in the census stations than recorded in the spontaneous calls. Spontaneous call method yielded the sighting of owls during 35 call sequences and 75% of sequences got response from the territory holders. Using the call playback method, 15 new responses were additionally recorded, where the presence of Brown hawk Owl were not recorded during the spontaneous call method. Hence, the broadcasting of conspecific call method can be considered as an important method for the census of Brown hawk Owl. Repeated call sequences in the same census point would not be a problem for the owls because habituation did not appear to be a problem for owl census (Gerhardt, 1991).

In nine sites, the Brown hawk Owl did not respond, but other two sympatric species responded (Barred jungle Owlet and Collard's scops Owl). Why the target individuals of Brown hawk Owl did not respond to the conspecific calls is not analysed here because of the inadequate data to support the hypothesis, "does the presence of collard's scops Owl and barred jungle Owl influence the vocal and response behaviour of Brown hawk Owl"? Similarly, Rohner (1997) reported a significant difference in the detection rate between the floaters and the pair male territorial owners in Great horned Owl, *Bubo virginianus*. Such an influence was also reported between the Barred Owl and the Boreal Owl, where Barred Owl influences the vocal behaviour of Boreal Owl (Takats & Holroyd, 1997).

Most of the spontaneous calls were heard during the pre breeding and breeding season than in other seasons when we tried the broadcast survey in a known site at day time, then it responded from the roost site twice but did not approach. And during the observation only one individual responded and the other partner was silent. Several factors are known to vary the vocal and response behaviour of owls and among them wind velocity, rainfall and social context are the major ones (Olsen *et al.*, 2002). As wind velocity increases, a "sound shadow" will be created, resulting in a marked attenuation of sound with greater distance. Attenuation also will be increased during rain because sound can be absorbed by water molecules and noise created by rainfall probably makes it more difficult to hear owl vocalization (Morrell *et al.*, 1991). Hence, the survey was not carried out during heavy rainfall and windy days to avoid bias in the data. The presence or absence of a breeding partner may also influence the rate of vocalization (Klatt & Ritchison, 1994).

The study showed that a viable population of Brown hawk Owl exists in the mid altitude moist forests of southern Western Ghats. This preference to the particular habitat makes the species an important indicator species of the mid altitude forest (Jayson, 2006). The quantity of canopy cover and tree density determines the approaching height and distance of Brown hawk Owl in response to conspecific calls.

Increasing canopy coverage, vertical complexity and tree density decreases the visibility of approaching owls. It can be concluded that (1) Brown hawk Owl responds to its broadcasted conspecific calls throughout the year with the highest detection rate during the breeding season (2) more studies are needed to understand why the response behaviour was different during other seasons of the year.

References

Appleby, B. M. & S. M. Redpath. 1997. Indicators of male quality in the hoots of Tawny owls, *Strix aluco*. *J. Raptor Res.* 31: 65-70.

Appleby, B. M., N. Yamaguchi, P. J. Johnson. & D. Mcdonald. 1999. Sex-specific territorial responses in Tawny owl, *Strix aluco*. *Ibis* 141: 91-99.

Debus, S. J. S. 1995. Surveys of large forest owls in Northern New South Wales: methodology, calling behaviour and owl responses. *Corella* 19: 38-50.

Debus, S. 1996. Vocal behaviour of the Southern Boobook, *Ninox novaeseelandiae* and other nocturnal birds. In: *Australian Raptor studies II. Birds of Australia Monograph*. Melbourne, Australia.

Galeotti, P. 1998. Correlates of hoot rate and structure in male Tawny owl, *Strix aluco*: implications for male rivalry and female mate choice. *J. Avian Biol.*, 29: 25-32.

Galeotti, P. & G. Pavan. 1993. Differential response of territorial Tawny owls (*Strix aluco*) to the hooting of neighbours and strangers. *Ibis* 135:300-304.

Galeotti, P., R. Sacchi & E. Perani. 1997. Cooperative defence and intrasexual aggression in Sops Owls (*Otus scops*): responses to playback of male and female calls. *J.Raptor Res.* 31: 353-357.

Gerhardt, R. P. 1991. Response of Mottled owls to broadcast of conspecific calls. *J.Field Ornithol.* 62 (2):239-244.

Ishtiaq, F & A. R. Rahmani. 2005. The forest owlet *Heteroglaux blewitti*: vocalization, breeding biology and conservation. *Ibis* 147:197-205.

Jayson, E. A. 2006. Ecology and behaviour of forest owls in the southern Western Ghats and developing habitat model for their conservation. Second year Annual report submitted to Ministry of Environment & Forests, Government of India, New Delhi.

Kavanagh, R. P. & P. Peake. 1993. Survey procedures for nocturnal forest birds: an evaluation of variability in census due to temporal factors, weather and technique. In: *Australian Raptor Studies* (Ed.) P.Olsen. Australasian Raptor Association, R.A.O.U., Melbourne. pp. 86-100

Klatt, P. H. & G. Ritchison. 1994. The effect of mate removal on the vocal behaviour and movement patterns of male and female Eastern Screech Owls. *Condor* 96:485-493.

Martinez, J. A. & I. Zuberogoitia. 2002. Factors affecting the vocal behaviour of eagle owls, *Bubo bubo*: Effects of sex and territorial status. *Ardeola* 49 (1): 1-9.

Morrell, T. E., R.H. Yahner, & W. L. Harkness. 1991. Factors affecting detection of Great horned owls by using broadcast vocalizations. *Wildl.Soc.Bull.* 19:481.

Mueller-dombois, D. & H. Ellenberg. 1974. *Aims and methods of vegetation ecology.* John Wiley, London. 547pp.

Neelakandan, K. K. 1980. The voice of juvenile brown hawk owl (*Ninox scutulata*). *J.Bombay. Nat .Soc.* 76 (2): 363-364.

Olsen, J., S. Trost, & G. Hayes. 2002. Vocalizations used by Southern Boobooks (*Ninox novaeseelandiae*) in the Australian capital territory. *Ecology and conservation of owls.* 305pp.

Ramanujam, M.C. 2004. Inter-specific intimidatory behaviour of adult Indian Eagle owl *Bubo bengalensis* (Franklin) in defence of their nestlings. *Zoo's Print* 19(2):1343-1345.

Rasmussen, P. C. & N. J. Collar. 1998. Identification, distribution and status of the Forest owlet, *Athene blewitti. Forktail* 14:43-51.

Takats, D. L. & G. L. Holroyd. 1997. Owl broadcast surveys in the foothills model forest, Alberta, Canada. *Biology and Conservation of Owls of the Northern Hemisphere,* Second International Symposium, pp. 347-362. Winnipeg, Manitoba, Canada.

Zuberogoitia, I. & L. F. Campos. 1998. Censusing owls in large areas: a comparison between methods. 1998. *Ardeola* 45: 47-53.

14

Effect of Environmental Factors and Typhoon's Flooding on the Population Dynamics of Benthic Fishes in Hapen Creek of Northern Taiwan

HEMA KRISHNAKUMAR HSIUNG LIANG
CHI-CHANG LIU AND YAO-SUNG LIN

ABSTRACT

Benthic fishes represent a significant component of riffle communities. Our 10 year study on two benthic fishes, *Crossostoma lacustre* and *Rhinogobius candidianus* which are sympatric in Hapen Creek, Northern Taiwan, reveals that these species exhibited mesohabitat overlaps but were segregated at the microhabitat levels. Both species preferred riffles and avoided rocks. Heavy flow (typhoons) had an inhibitory effect on the fish densities. The fishes may not be able to swim against the strong currents and might get carried away downstream. The recovery time (number of days after a peak, when the densities were restored to normalcy) for *C. lacustre* was = 13 days and for *R. candidianus*, it was = 16 days. Recovery was faster at the downstream (~13-15 days) as opposed to upstream (~14-16 days). Direct estimation of the linkage between the hydrological or geomorphological variables and the well being of fish populations is one of the first steps in formulating an efficient, effective conservation plan or restoration strategy to preserve these species for the future.

INTRODUCTION

Hydrological and geomorphological conditions of streams are highly variable and dynamic, thus providing diverse habitats for fishes and other aquatic organisms. These characteristics have strong effects on the abundance and distribution of fishes. The association between fish and habitats also provides fundamental information for stream restoration and management. Several models have been developed, which link the role of habitat variables, individually and collectively, to the abundance of fishes (Chuang *et al.*, 2004).

Sympatric species which are congeneric or ecologically similar are often considered as competing for resources (Winston, 1995). However, studies have shown that such stream fishes often exhibit habitat shifts, suggesting that their habitat segregation in sympatry is a consequence of interspecific interactions (Sone *et al.*, 2001). Resource partitioning in a community can occur along the trophic, spatial and temporal dimensions, and fish assemblages in most habitats segregate along spatial and trophic levels (Ross, 1986).

Water flow is one of the most important factors influencing fish abundance in streams (Sone *et al.*, 2001). Floods have a major impact on the survival of populations, reproduction, spawning and the dispersal of larvae and are dependent on these events and are synchronized with them (Keith, 2003). Floods also have a major role on the alteration of habitat structures. In Taiwan, heavy flow during typhoons can alter the fish distribution in mountain streams. Typhoons bring in torrential rainfalls and are among the most important hydrological characteristics which affect headwaters or mountain streams (Hsia & Hwong, 1999). Several studies (Lee *et al.*, 1998; Han *et al.*, 2000, Chuang *et al.*, 2004) focus on the spatial distribution and habitat use of freshwater fishes in Taiwan, but very few on the role of water flow especially flooding effect on the population of benthic stream fishes.

Crossostoma lacustre (Balitoridae) and *Rhinogobius candidianus* (Gobiidae) are endemic fish species of Taiwan. They are benthic and sympatric in many mountain streams of northern and central Taiwan (Wang & Shao, 1997; Chang *et al.*, 1999; Chen & Fang, 1999) and also in Hapen Creek, a protected stream in the LTER site-Fushan

Experimental Forest, Northern Taiwan (Chuang *et al.*, 2004). The purpose of this study was to describe the population dynamics of *C. lacustre* and *R. candidianus* in Hapen Creek. The specific objectives were to (1) examine the differences in the densities of *C. lacustre* and *R. candidianus* along the temporal and spatial scale, (2) estimate the role of environmental variables on the benthic fish densities (3) determine the influence of flow on fish densities and the effects of high flow caused by typhoon, and (4) estimate the time required for post-typhoon recovery of the two species.

MATERIAL & METHODS

Study area and sampling sites: The study was conducted in Hapen creek, a headwater tributary of the Nanshih Stream at elevations of 500-1200 m. It is a natural, well-protected mountain creek in the Fushan Experimental Forest, Ilan, Northern Taiwan (Fig.1). Four seasons have been identified here: spring (March-May), summer (June-August), fall (September-November) and winter (December-February).

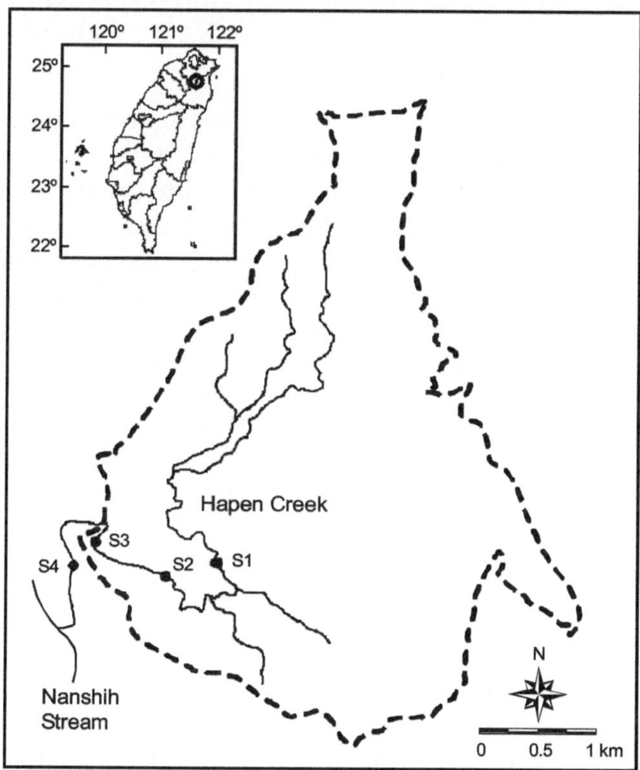

Fig. 1. Map of Hapen creek with the four study stations

The mean monthly precipitation in Hapen creek during 1996-2001 was 2251.1 mm and the mean annual discharge was $0.395 m^3 s^{-1}$, with a mean monthly discharge ranging between $0.003 m^3 s^{-1}$ and $0.311 m^3 s^{-1}$. The mean monthly air temperature ranged from 8.6°C to 30.3°C.

The riparian zone of Hapen Creek was covered primarily by natural broad leaf forests. There are six fish species in this stream: *Varicorhinus barbatulus, Candidia barbatus, Crossostoma lacustre, Rhinogobius candidianus, Cobitus sinensis* and *Acrossocheilus paradoxus*, of which, *V. barbatulus* and *C. barbatus* are the most abundant ones (Chang *et al.*, 1998).

Four study sites, S1, S2, S3 and S4 (from upstream to downstream) were selected, each with four habitat units (mesohabitats): two riffles and two pools. Riffles were identified as a section of the stream where the water surface has visible standing or breaking anti-waves (Gelwick, 1990). The distances of S1, S2, S3 and S4 to the confluence of the Nanshih stream (tributary of the Tanshui river) is 4.2 Km, 2.3 Km, 1.6 Km and 0.7 Km, respectively. The average gradient of the study area is 17.5 m/Km (Chang *et al.*, 1998).

Fish sampling: Bimonthly fish sampling was conducted at each study site from February 1996 to December 2003. Fishes were sampled by electrofishing for a 15-minute period from downstream to upstream as a zig-zag pattern at each habitat unit. A battery-powered, backpack-mounted electrofisher (150-300V, 1A pulsed DC) was used. Two field assistants helped to collect the stunned fishes with dip nets. Collected fishes were identified to their species levels and their total lengths (TL) were measured in the field.

Environmental Measurements: Hydrological variables and habitat conditions were measured for each riffle and pool using methods described by Tsao (1995) immediately after fish sampling. Six parameters (microhabitat conditions), including stream width, water depth, substrate, water velocity, in stream cover and canopy were measured along three permanent transects perpendicular to the current flow at each habitat unit.

The current flow on a sampling day at each habitat unit was calculated as: Flow (m^3/sec) = velocity (m/sec) * width (m) * depth (m). The daily water flow during the entire study period was obtained from the Fushan Gauging Station. Current flow in a stream is never uniform. Even for a one year period, a stream can be subjected to various hydrological conditions like typhoon, drought or heavy rainfall and the current flow depends on these conditions. Hence, the mean flow for each year was calculated separately and the entire flow data was grouped into five categories based on the hydrological conditions observed during a sampling interval (like the hydrological condition in the stream between first sampling date (1st February 1996) to the next sampling date (1st April 1996), which is a period of ~ 2 months). The flow categories are:

A. Flow around mean level during entire period;

B. Flow around mean level followed by a high peak just before the next sampling;

C. A high peak occurs, followed by flow at below ½ the mean level;

D. Flow below ½ mean level followed by a high peak just before the next sampling;

E. Flow below ½ means level during entire period.

Data analysis: Fish density was calculated by dividing the number of fishes counted by sampled area of each habitat (stream length * mean stream width). Paired t-tests were used to compare fish densities between pools and riffles. Analysis of variance (ANOVA) was used to examine the spatial (site), temporal and seasonal variation in density. We used factor analysis to identify the interactions among the major environmental variables in Hapen creek. Subsequently, we used stepwise regression analyses to investigate the relationship of environmental factors with fish densities. We also individually examined the relationship between fish densities and flow levels using Spearman's correlation analysis. Multi-dimensional scaling ordination (MDS) of fish density variability was performed for each flow category at an upstream (site 1) and at a downstream site (site 4). The recovery time was determined by counting the number of days after a peak, when the fish densities were restored to their pre-sampling levels.

RESULTS

The hydrological variables in Hapen Creek fluctuate highly. During the study period, water flow ranged from 0 m^3/sec to 4.85 m^3/sec with an average of 0.37±0.012 m^3/sec and the current velocities ranged from 0 m/sec to 1.32 m/sec. The water depth in the stream ranged from 0.02 m to 1.30 m. The stream width varied from 1.7 m to 15.0 m. During the low flow months, water level drops significantly in many riffles, although it flows subterraneously.

Variations in density along the temporal and spatial scale: A total of 6881 samples were collected. We captured a total of 4022 *C. lacustre* during the study period: 3068 from riffles with a mean of 383.5±245.7 and 954 from pools with a mean of 119.3±134.1. The density was significantly higher in riffles than in pools (t = 2.4, df = 7, p = 0.05). However, there were no significant spatial or temporal variations in the density of *C. lacustre* within each sampling site (p > 0.05) (Table 1).

A total of 2859 *R. candidianus* were captured: 2405 from riffles with a mean of 300.6±133.5 and 454 from pools with a mean of 56.8±26.0. The density was significantly higher in riffles than in pools (t = 5.6, df = 7, p = 0.001). The density of *R. candidianus* did not vary spatially or temporally in pools. However, in riffles, spatial variations are significantly higher at the upstream sites than at the lower stream section (Table 1). No seasonal variations in the densities of the benthic fishes at Hapen creek were found (*C. lacustre*: $F_{3, 29} = 1.28$, p = 0.3; *R. candidianus*: $F_{3, 29} = 0.24$, p = 0.87).

Effect of habitat variables on fish densities: The factor analyses of the environmental variables show that 65.3% of the variation was retained in 3 factors (Table 2). Factor 1 (rocks) was mainly composed of cobbles and some large boulders and retained 20.8% of the variation. Factor 2 (rocky depths) comprised of some large and small boulders and water depth and retained 38.1% of the variation. Factor 3 (silty gravel) was composed of pebbles, cobbles and sand and retained 6.4% of the variation. Rocks (F1) and rocky depths (F2) were negatively correlated with fish densities of both species (*C. lacustre*- F1: r = -0.251; F2: r = -0.257; *R. candidianus*- F1: r = -0.346; F2: r = -0.329) and silty gravel (F3) was positively correlated in *C. lacustre*- F3: r = 0.176; *R. candidianus*- F3: r = 0.023). Thus, benthic fishes in Hapen creek avoid rocks and prefer shallow water areas with less larger substrates like pebbles, cobbles and silt. Moreover, the stepwise multiple regression analyses with these factors and the environmental variables show that *C. lacustre* densities (CLd) had significant positive correlations with silty gravel (F3), but negative correlations with rocky depths (F2). The relationships are expressed by the equation:

$$CLd = 24.072 + 0.385\ F3 - 0.220\ F2$$
$$R^2 = 0.139,\ df = 147$$

Table 1: Results of the statistical tests performed on the two benthic fish species P: Pool, R: Riffles; F1: 1 way ANOVA. Variables tested include: density of fish species and stations; F2: 2 way ANOVA. Variables tested include: density of fish species, stations and years.

Species	Habitat	Tests	*F*	*df*	p
Crossostoma lacustre	P	F1	0.04	1, 23	0.853
		F2	0.70	2, 22	0.519
	R	F1	0.90	1, 29	0.344
		F2	1.00	2, 28	0.389
Rhinogobius candidianus	P	F1	0.01	1, 22	0.941
		F2	0.50	2, 21	0.615
	R	F1	10.10	1, 30	0.003
		F2	4.90	2, 29	0.015

Table 2: Model summary of factor analysis of the environmental variables in Hapen creek, and correlation of the factors and its constituents with fish density. 65.3% of the variances are retained by 3 factors: Rocks (cobbles and large boulders), Rocky depths (large and small boulders and water depth) and Silty gravel (pebbles, cobbles and sand).

Factors *Correlation with fish density* (N = 155)	**% Variance explained**	*r*	*t*	*p*
Factor 1 (Rocks)				
C. lacustre	20.8	-0.251	-3.203	0.002
R. candidianus		-0.346	-4.569	0.000
Factor 2 (Rocky depths)				
C. lacustre	38.1	-0.257	-3.294	0.001
R. candidianus		-0.329	-4.313	0.000
Factor 3 (Silty gravel)				
C. lacustre	6.4	0.176	2.216	0.028
R. candidianus		0.023	0.282	0.778

The R^2 value suggests that 13.9% of variance of *C. lacustre* densities were significant and can be explained by the substrate factors (silty gravel and rocky depths). Rocky depths were responsible for 9.0% and silty gravel for 4.9% of the variances (Table 3). Thus, *C. lacustre* prefers shallow water in riffles with silt, or pebbles and avoids rocks. *R. candidianus* densities (RBd) showed only significant negative relationships with stream width (W), and rocks (F1). The relationships are expressed by the equation:

$$RBd = 34.563 - 1.178\ W - 0.248\ F1$$
$$R^2 = 0.172,\ df = 147$$

The R^2 value suggests that 17.2% of variance of fish densities were significant and can be explained by a hydrological factor, stream width and by the substrate factor (rocks). Rocks were accountable for 10.9%, and stream width for 6.3% of the variances (Table 3). Thus, *R. candidianus* prefers narrow stream areas and avoids rocks.

Table 3: Model summary of stepwise multiple regressions of densities of *C. lacustre* and *R. candidianus* with environmental variables. Only significant variables have been shown.

Variables	R^2	Regression coefficients	Standard errors	*F*	*P*
C. lacustre					
(Total R^2 = 0.139, N = 151)					
Silty gravel (F3)	0.049	0.336	0.121	7.733	0.006
Rocky depths (F2)	0.090	-0.205	0.053	14.779	0.000
R. candidianus					
(Total R^2 = 0.172, N = 151)					
Stream width	0.063	-2.211	0.704	9.873	0.002
Rocks (F1)	0.109	-0.300	0.071	18.088	0.000

Water flow and fish densities: During the study period, we recorded peaks in water flow (>0.3 m^3/sec) during June-November. During August-November 85.19 % of the high flow was concentrated. The flow levels were significantly higher during June-November than during December-May ($t = 10.5$, $df = 5$, $p < 0.001$; Fig. 2).

We also observed that immediately after a flooding event, there was a significant decrease in densities of *C. lacustre* and *R. candidianus* and within a few days, the densities were restored to their pre-sampling levels (Tables 4 & 5). Strong negative correlations between water flow and fish densities were observed for both benthic species (*C. lacustre*: $r = -0.677$, $P < 0.01$; *R. candidianus*: $r = -0.632$, $P < 0.01$; Fig. 3).

The MDS analyses of fish densities at an upstream and a downstream site illustrate the influence of flow variability on fish density

(Fig. 4). There is a clear separation of flow categories C and E, which deals with post-peak fish density recovery and stability in the upstream site (site 1). However at the downstream site (site 4), the flow categories C and E are not separated from A, B & D categories, indicating that the flow categories are not reliable as in the other three sites. We further tested the differences in the water depth, stream width and current velocity at site 1 and site 4 and found that there were no significant differences in current velocity between the two sites but, water depth and stream width were higher in site 4 (depth: $t = 6.327$, $df = 38$, $P < 0.001$; width: $t = 4.249$, $df = 38$, $P < 0.001$. Hence, it is evident that the flow in site 4 was higher than in site 1. This clearly reveals that the recovery time of fishes varies among the upstream and downstream sites. We also performed a separate MDS analysis on site 1 and 4 separately for both fish species to identify relations among units, but the ordination was nearly identical to the analysis using all data. The recovery time of fish density were estimated. In *C. lacustre* it was estimated to require at least 13 days and for *R. candidianus* it was estimated at 16 days (Fig. 5). Recovery of fish density at the upstream site was slower (~14-16 days) than at the downstream (~13-15 days).

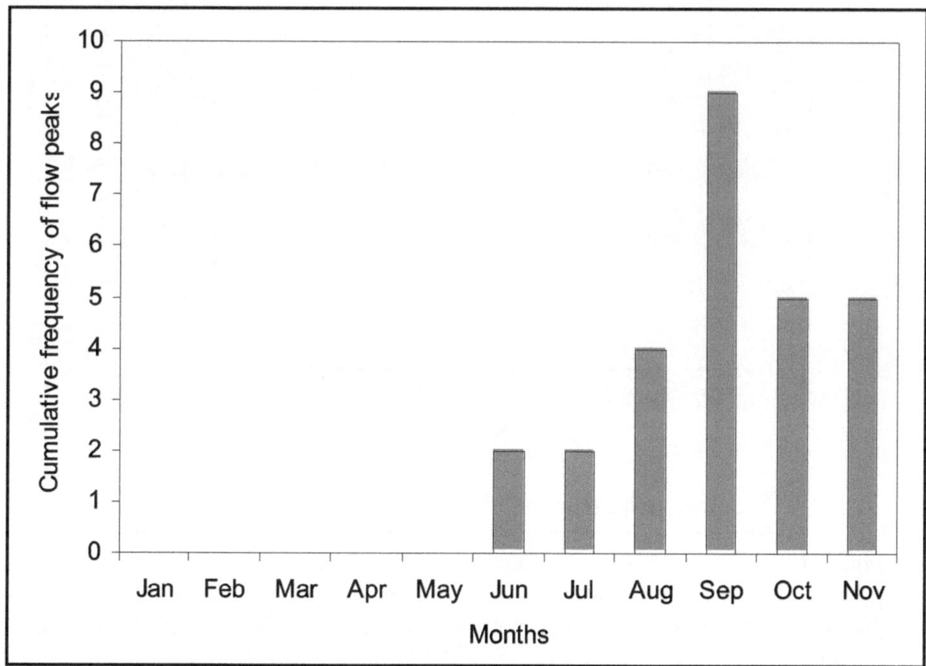

Fig. 2. Cumulative frequency of high peaks in Hapen creek during the study period (1996- 2003). Flow > 0.3 cms has been identified as a peak.

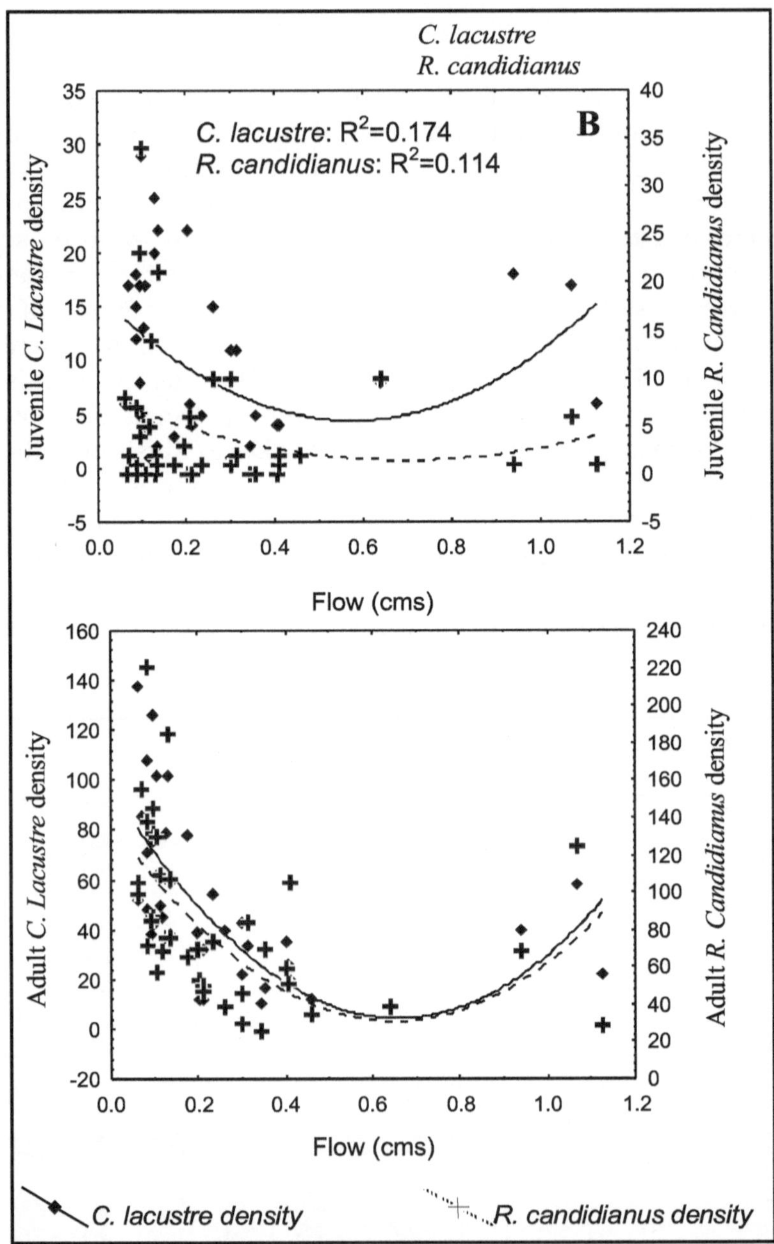

Fig. 3. Effect of water flow on fish densities. Flow >0.3 has been considered as a peak. (A) Juvenile densities of *C. lacustre* (Total length < 50 mm) and *R. candidianus* (Total length < 35 mm). *C. lacustre* juvenile density = 16.161 – 40.925x + 35.546x^2; *R. candidianus* juvenile density = 8.320 – 19.809x + 14.216x^2. (B) Adult densities of *C. lacustre* and *R. candidianus*. *C. lacustre* adult density = 97.557 – 283.516x + 215.595x^2 ; *R. candidianus* adult density = 138.274 – 330.184x + 253.983x^2

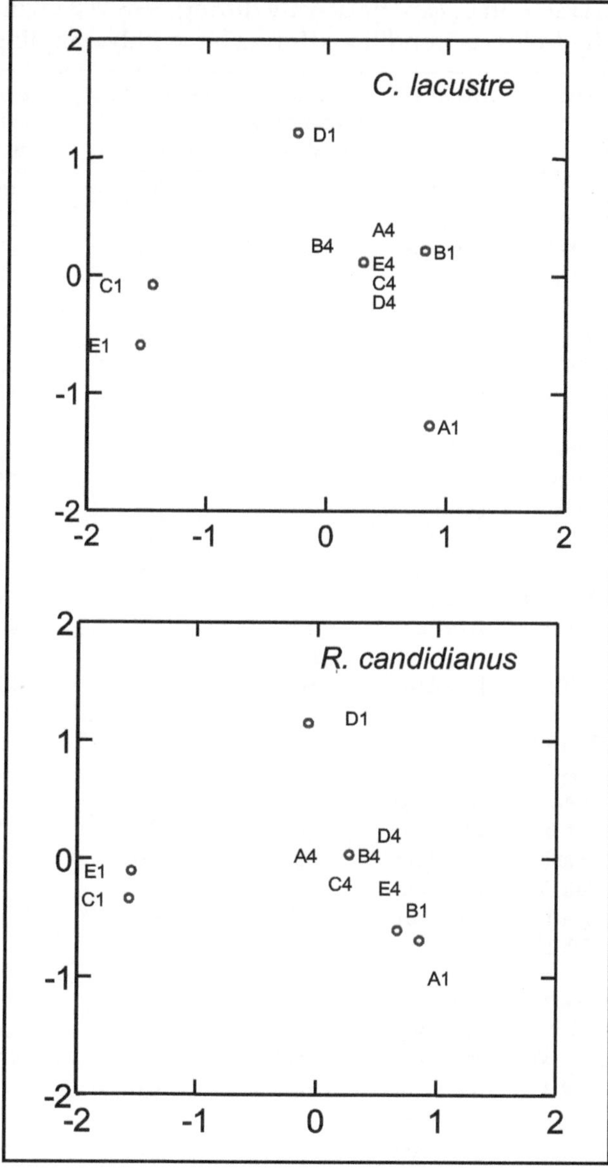

Fig. 4. MDS ordination of fish density variability after bimonthly intervals during the study period (1996-2002). In *C. lacustre*, stress = 0.05 and in *R. candidianus*, stress = 0.028. The stress values for the two species indicate that the ordination provides excellent representation of the data. Each alphanumeric data point represents the density variability after the occurrence of each flow category (A-E; see Tables 4 & 5); 1 represents the upstream site and 4 represent the downstream site. The MDS shows a site specific reaction to the flow pattern. Site 4 is similar for the 5 hydrological categories for both the species.

Table 4: Variations in *C. lacustre* density during 1996-2002 with the occurrence of different hydrological conditions during bimonthly sampling intervals.

Flow category	Year	Months	Density before flow condition	Density after flow condition	Days to the last peak	% increase/ decrease in density
A	1996	Apr-Jun	34	38	18	11.77
	1997	Aug-Oct	52	155	39	66.45
	1998	Feb-Apr	80	157	26	96.25
	2000	Feb-Apr	131	86	0	-34.35
B	1996	Aug-Oct	75	49	0	-34.67
	1997	Aug-Oct	107	31	2	-245.16
	2000-2001	Dec-Feb	107	46	10	-57.01
C	1996	Feb-Apr	60	34	14	-43.33
	1996	Jun-Aug	38	75	16	97.37
	1996	Oct-Dec	49	115	31	134.69
	1998	Apr-Jun	157	66	14	-57.96
	1999	Oct-Dec	30	69	24	130.00
	1999-2000	Dec-Feb	69	131	14	89.39
	2000	Apr-Jun	86	223	16	159.30
	2000	Aug-Oct	137	155	14	13.14
	2001	Jun-Aug	69	54	11	-21.74
D	1996-1997	Dec-Feb	115	58	0	-49.57
	1997	Apr-Jun	138	57	0	-58.70
	1997	Jun-Aug	57	52	0	-8.77
	1998	Oct-Dec	31	26	0	-16.13
	2000	Jun-Aug	223	137	2	-38.57
	2000	Oct-Dec	155	107	3	-30.97
	2001	Apr-Jun	203	69	1	-194.20
	2001	Aug-Oct	54	35	8	-54.29
E	1997	Feb-Apr	58	138	57	137.93
	1997	Oct-Dec	155	76	96	-50.97
	1997-1998	Dec-Feb	76	80	159	5.26
	1998	Jun-Aug	66	107	78	62.12
	2001	Feb-Apr	46	203	77	341.30
	2001	Oct-Dec	35	73	54	108.57
	2001-2002	Dec-Feb	73	135	78	84.93
	2002	Feb-Apr	135	144	136	6.67
	2002	Apr-Jun	144	104	196	-27.78

- Flow categories: A - Flow around mean level during entire period; B - Flow around mean level followed by a high peak just before the next sampling; C - A high peak followed by flow at below ½ the mean level; D - Flow below ½ mean level followed by a high peak; E - Flow below ½ mean level during entire period.

** Mean flow levels were calculated for each year separately

Table 5. Variations in *R. candidianus* density during 1996-2002 with the occurrence of different hydrological conditions during bimonthly sampling intervals.

Flow category	Year	Months	Density before flow condition	Density after flow condition	Days to the last peak	% increase/ decrease in density
A	1996	Apr-Jun	15	10	18	-33.33
	1997	Aug-Oct	27	134	39	79.85
	1998	Feb-Apr	99	117	26	18.18
	2000	Feb-Apr	127	103	0	-18.90
B	1996	Aug-Oct	55	18	0	-67.27
	1997	Aug-Oct	144	23	2	-526.09
	2000-2001	Dec-Feb	24	19	10	-20.83
C	1996	Feb-Apr	40	15	14	-62.50
	1996	Jun-Aug	10	55	16	450.00
	1996	Oct-Dec	18	59	31	227.78
	1998	Apr-Jun	117	81	14	-30.77
	1999	Oct-Dec	28	58	24	107.14
	1999-2000	Dec-Feb	58	74	14	27.59
	2000	Apr-Jun	45	63	16	40.00
	2000	Aug-Oct	74	45	14	-39.19
	2001	Jun-Aug	22	44	11	100.00
D	1996-1997	Dec-Feb	59	54	0	-8.48
	1997	Apr-Jun	105	39	0	-62.86
	1997	Jun-Aug	39	27	0	-30.77
	1998	Oct-Dec	23	12	0	-47.83
	2000	Jun-Aug	63	127	2	101.59
	2000	Oct-Dec	103	24	3	-76.70
	2001	Apr-Jun	121	22	1	-81.82
	2001	Aug-Oct	44	13	8	-70.75
E	1997	Feb-Apr	54	105	57	94.44
	1997	Oct-Dec	134	120	96	-10.45
	1997-1998	Dec-Feb	120	99	159	-17.50
	1998	Jun-Aug	81	144	78	77.78
	2001	Feb-Apr	79	58	77	-26.58
	2001	Oct-Dec	13	41	54	215.38
	2001-2002	Dec-Feb	19	121	78	536.84
	2002	Feb-Apr	41	75	136	82.93
	2002	Apr-Jun	75	79	196	5.33

- Flow categories: A - Flow around mean level during entire period; B - Flow around mean level followed by a high peak just before the next sampling; C - A high peak followed by flow at below ½ the mean level; D - Flow below ½ mean level followed by a high peak; E - Flow below ½ mean level during entire period.

** Mean flow levels were calculated for each year separately

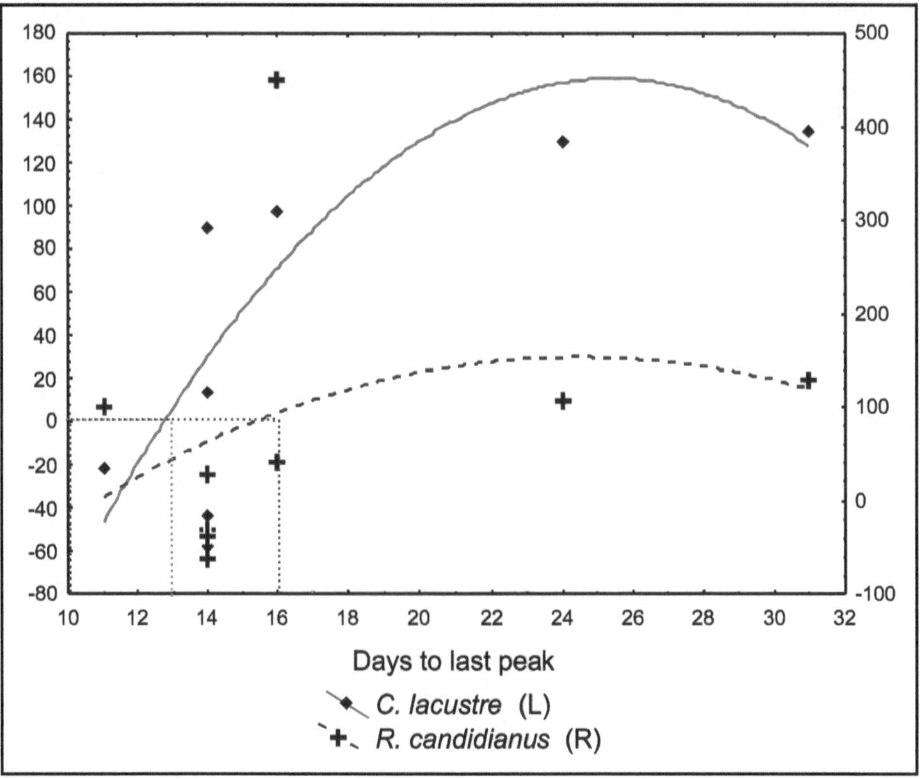

Fig. 5. Figure showing the recovery of fishes after a peak. The samples used were those in flow category 'C' where a high peak was followed by flow at below ½ the mean level. Recovery time for *C. lacustre* was = 13 days and for *R. candidianus* it was = 16 days

DISCUSSION

Habitat use : Benthic fishes have been recognized to represent a significant component of the riffle communities (Gray & Stauffer, 1999). The study shows that *C. lacustre* and *R. candidianus* preferred riffles, exhibiting mesohabitat overlaps but segregated at the microhabitat levels. Preference for riffles may be attributed to their diet proportion on aquatic insects (Chuang *et al.*, 2004). Generally, invertebrates are prevalent in stony riffles than in pools (Allan, 1995). *C. lacustre* is an omnivorous feeder, feeding mainly on benthic algae and aquatic insects (Huang, 2002) and has no preference for sites upstream or downstream. It preferred shallow water in riffles with silty gravel and avoided rocks. In contrast, *R. candidianus* mainly feeds on aquatic insects and macro invertebrates (such as freshwater crabs) hidden in underwater crevices (Huang, 2002) and prefers sites which are upstream where the flow is slower. This conforms the observation by Sone *et al.* (2001) wherein *Rhinogobius* sp. were found to prefer lower bottom velocity in a fast current. *R. candidianus* is not a fast swimmer compared to *C. lacustre.* Another possible explanation is the amount of plant cover on the banks. *Rhinogobius* density was found to be highest when canopy (plant cover) was poorest (Keith, 2003). Canopy has an inverse relationship with the algal growth and in turn on the aquatic insects (primary consumers) on which these gobies feed (Inoue & Nakano, 2001). In Hapen Creek, the canopy levels are lower upstream (54.8%) as opposed to downstream (79.2%). *R. candidianus* on the other hand, preferred silty shallow water areas and avoided rocks. Dudgeon (1987) suggests that the differences in the microhabitat utilization among homalopterids and gobies result from their morphological variations rather than competition. This may also partially contribute to the segregations at the microhabitat level between the two benthic fishes in Hapen creek.

Fish Response to flooding : The mean yearly water flow over a 10 year period, ranged between 0.15-0.24 m^3/sec. A flow >0.3 m^3/sec was categorized as a peak and it always coincided with a typhoon. Typhoons greatly increase the erosive power of streams (Tew *et al.*, 2002), and can cause landslides and debris flows in mountain regions. Heavy water flow had an inhibitory effect on the density of both species of fishes. There was always a drastic drop in fish densities immediately after a typhoon. The benthic fishes avoid boulders which form underwater crevices, where they can escape from the strong currents. Instead, they prefer silty gravel and perhaps are not able to swim against the currents and might get carried away downstream. Vulnerability to flood discharge has been recorded in the Japanese fluvial sculpin, *Cottux pollux* (Natsumeda, 2003) and also in many stream fishes in South America (Almiron *et al.*, 2000). In Hapen creek, after a few days, the densities were restored to normal condition. This is totally contradictory to the results obtained by Tew *et al.*, (2002) showing the effect of a catastrophic typhoon, Herb (during July 1996) on the fish abundance in Qi-Shan stream in south Taiwan, where the reported recovery time was between 14-17 months. However, our data for that particular typhoon (flow: 0.82-0.91m^3/sec) show that after just 3 days, the flow levels dipped to below mean level for that year (0.21m^3/sec) and on our next sampling day (August 17, 1996), the fishes had already recovered (recovery time: < 16 days). A possible explanation is that this typhoon was more severe in southern Taiwan than in the Northern part as is evident from the precipitation levels (Alishan, Southern Taiwan: 1094.5 mm; Ilan, Northern Taiwan: 347.4 mm) and from the damage caused. The roads to Qi-Shan stream were severely damaged, making it inaccessible for fish sampling for about a year. Secondly, the frequency of our sampling was ~60 days but in the other study, the post-typhoon sampling was conducted after one year. During a period of one year, a stream can be subjected to various hydrological or geomorphological conditions like typhoons, drought, heavy rainfall and earthquakes (Fang *et al.*, 2002). Our data shows that even with a ~60 day sampling period, the hydrological conditions in the stream varied greatly, forcing us to classify them based on the flow levels during the 60 day interval. Thirdly, the gradient of the flow can also play a role in the recovery. Post-typhoon fish recovery in Hapen creek was faster at the downstream than at the upstream. In fishes, momentary swimming ability against water currents is positively correlated with the physical condition (Kodric-Brown & Nicoletto, 1993). Immediately after a typhoon, the fishes are physically weak and have a lower swimming performance (personal observation, unpublished) and swimming against fast currents may require considerable body energy than usual.

CONCLUSION

Few studies have assessed the role of water flow on benthic fishes and the post-typhoon recovery as we have analyzed here. Our data reveals that benthic fishes in Hapen creek exhibit mesohabitat overlap but are segregated at the microhabitat levels. Differences in the microhabitat utilization result from morphological variations among sympatric species rather than competition (Chuang *et al.*, 2004). Heavy flow has an inhibitory effect on benthic fish density. The post-typhoon recovery time for benthic fishes in Hapen creek was ~13-16 days with faster recovery at the downstream.

This is the first report on the recovery of fish density after flooding events in an LTER site in Taiwan. The effects of devastating typhoons are poorly documented, mainly because investigators lack information on the pre-disturbed conditions (Tew *et al.*, 2002). We find it difficult to explain the speedy recovery due to limited studies and hence, a dearth of empirical evidences to support our observations. Also, there is a limitation in our understanding the upstream-downstream effects. To completely reveal the stream system dynamics, emphasis must be made on the spatial and temporal relationships, food provisioning, and predator-prey interactions for fish species within the stream. Invertebrate abundance (numbers and biomass) has been established as one of the most important determinants of habitat selection for some benthic fishes (Sone *et al.*, 2001). Additionally, our stepwise multiple regression analysis showing the relationship between environmental variables and fish density has low R^2 value (0.13-0.17) for both benthic species, indicating that the explanation power of the included variable has a limited ability in explaining the variability of fish density. Certainly, part of this observation could be the result of a large sample size. Understanding the linkage between the environmental variables and the well being of fish populations is one of the first steps in formulating an efficient, effective conservation plan or restoration strategy to preserve these species for the future (Keith, 2003). Long-term stream monitoring can provide further insight into potential ecosystem changes that occur as a result of changes in the geomorphology of a stream.

Acknowledgements

We are grateful to the Taiwan Forestry Research Institute for providing facilities and accommodations at Fushan Research Center. We also thank the research assistants and students of the Biodiversity Research Lab, Department of Life Science, National Taiwan University for helping in field work. This work was supported by grants from the National Science Council of the R.O.C. (Project number: NSC 93-2621-B002-003) during the study period.

References

Allan, J. D. 1995. *Stream Ecology, Structure and Function of Running Waters.* Chapman and Hall, London. 388 pp

Almiron, A. E., M. L Garcia, R. C. Menni, L. C. Protogino & L. C. Solari. 2000. Fish ecology of a seasonal lowland stream in temperate South America. *Mar. Fresh. Res.* 51(3): 265-274.

Chang, S. H., P. F. Lee & Y. S. Lin. 1998. Reproductive biology of *Varicorhinus barbatulus* in Hapen creek, Fushan, Taipei County, Taiwan. *Ann. Taiwan Mus.* 41: 53-69.

Chang, M. S., Y. S. Lin & L. C. Chuang. 1999. Effect of dams on fish assemblages of the Tachia river, Taiwan. *Acta Zool.* Taiwan 10: 77-90.

Chen, I. S. & L. S. Fang. 1999. *The freshwater and estuarine fishes of Taiwan.* National Museum of marine biology and aquarium, Pingtung. pp. 287.

Chuang, L. C., S. H. Liang & Y. S. Lin. 2004. Habitat use of two benthic fishes, *Crossostoma lacustre* and *Rhinogobius candidianus* in Hapen creek of Northern Taiwan. *Taiwania* 49(3): 166-174.

Dudgeon, D. 1987. Niche specificities of four fish species (Homalopteridae, Cobitidae and Gobiidae) in a Hong Kong forest stream. *Arch. Hydrobiol.* 108: 349-364.

Fang, L. S., I. S. Chen, K. S. Tew, C. C. Han, T. F. Lee & I. M. Chen. 2002. Impact of the 1999 earthquake on mountain stream fishes in Taiwan. *Fisher. Sci.* 68 (2): 446-448

Gelwick, F.P.. 1990. Longitudinal and temporal comparisons of riffle and pool assemblages in a northeastern Oklahoma Ozark stream. *Copeia* 1990: 1072-1082.

Gray, E. V. S. & J. R. Stauffer Jr. 1999. Comparative microhabitat use of ecologically similar benthic fishes. *Environ. Biol. Fishes* 56: 443-453.

Han, C. C., K. S. Tew, I. S. Chen, L. Y. Su & F. S. Fang. 2000. Environmental biology of an endemic cyprinid, *Varicorhinus alticorpus*, in a subtropical mountain stream of Taiwan. *Environ. Biol. Fishes* 59: 153-161.

Hsia, Y.J. & J.L. Hwong. 1999. Hydrological characteristics of Fushan experimental forest. *Q. J. Chin. For.* 32: 39-51.

Huang, I.Y. 2002. *Stream food web of a subtropical woodland headwater (Hapen Creek): insights from stable carbon and nitrogen isotopes.* MS thesis, National Taiwan University, Taipei, Taiwan. (in Chinese).

Inoue, M & S. Nakano. 2001. Fish abundance and habitat relationships in forest and grassland streams, northern Hokkaido, Japan. *Ecol. Res.* 16 (2): 233-247.

Keith, P. 2003. Biology and ecology of amphidromous Gobiidae of Indo-Pacific and Caribbean regions. *J. Fish Biol.* 63: 831-847.

Kodric-brown, A & P. F. Nicoletto. 1993. The relationship between physical condition and social-status in pupfish, *Cuprinodon pecosensis. Anim. Behav.* 46 (6): 1234-1236.

Lee, T. W., J. J. Choue, W. L. Lin & S. L. Yu. 1998. Environmental factors affecting the distribution of the deep-body shovelnose minnow, *Varicorhinus alticorpus*, in the Bay-nan river, Taiwan. *Q. J. Chin. For.* 31: 219-225.

Natsumeda, T. 2003. Effects of a severe flood on the movements of Japanese fluvial sculpin. *Environ. Biol. Fishes* 68 (4): 417-424.

Ross, S. T. 1986. Resource partitioning in fish assemblages; a review of field studies. *Copeia* 1986: 352-388.

Sone, S., M. Inoue & Y. Yanagisawa. 2001. Habitat use and diet of two stream gobies of the genus *Rhinogobius* in south-western Shikoku, Japan. *Ecol. Res.* 16: 205-219.

Tew, K. S., C. C. Han, W. R. Chou & L. S. Fang. 2002. Habitat and fish fauna structure in a subtropical mountain stream in Taiwan before and after a catastrophic typhoon. *Environ. Biol. Fishes* 65 (4): 457-462.

Wang, S. C. & K. T. Shao. 1997. Species composition and distribution of freshwater fishes in the rivers of Keelung region. *Chin. Biosci.* 40: 30-32.

Winston, M. R. 1995. Co-occurrence of morphologically similar species of stream fishes. *Am. Nat.* 145: 527-545.

15

Leaf-selection and Tent Construction by the Indian Short-nosed Fruit Bat, *Cynopterus sphinx* (Chiroptera: Pteropodidae) in the Palm, *Borassus flabellifer*

RAJASEKAR, R. AND K. SRIPATHI

ABSTRACT

The Indian short-nosed fruit bat, *Cynopterus sphinx* is one among the three paleotropical species known to make tents. Once a tent is made, male typically defends it against other competing males and also defends the females who join his tent. While taking into account the amount of time and energy invested in tent making, it has been proposed that tent making bats must select leaves with specific morphology so as to confer maximum benefits in terms of protection and stability. Consistently, a number of studies on neotropical tent making bats have shown that bats preferred leaves with specific characteristics. Our study in the palm *Borassus flabellifer* shows that *Cynopterus sphinx* preferred larger leaves with thicker rachis and leaves at specific angle. Bat-preferred leaves were found without any vegetation at the immediate vicinity of the roost opening. Our study also shows that males alone participate in tent-construction.

□□□

16

Bat Pollination Syndromes in Few Species of Plants

THIRUCHENTHIL NATHAN, P., IRULAN
AND G. MARIMUTHU

ABSTRACT

Folivorous bats consume flowers and fruits, effecting seed dispersal and pollination. Adaptation of plants according to the type of pollinator is called pollination syndrome. Some of the adaptations of bat-pollinated syndromes are nocturnal anthesis, odour emission and presentation of rewards to the pollinators. We studied the bat-pollinated plants such as *Parkia biglandulosa, Bassia latifolia, Ceiba pentandra* and *Kigelia pinnata.* Flowers of these plants emit strong odour and are well exposed to bats. Except *C. pentandra,* which gives its entire corolla as a reward, the other plants produce copious quantities of nectar for the visiting bats. These plants open their flowers in synchrony with the nocturnal activity of bats. The plants also produce large number of flowers each night for a shorter period of days showing big bang strategy to attract large number of bats. The bat visitations of these plants are synchronized with the time of anthesis and the maximum production of nectar. The pollen attached to the fur of these flying mammals aids cross-pollination. The adaptations of these plants for bat pollination imply the co-evolution of bats and plants pollinated by them.

17

Movement of Coconut Eriophyid Mite, *Aceria guerreronis* Keifer on Nut Surface: A Behavioural Study

MALLIK, B., J. JAYAPPA AND H. GURUPRASAD

ABSTRACT

The coconut eriophyid mite, *Aceria guerreronis* Keifer is a serious pest of coconut in Kerala, Tamil Nadu, Karnataka, Andhra Pradesh, Orissa, West Bengal, Maharashtra and Gujarat. The mites reside underneath the perianth of nuts and feed on the soft meristematic tissues of the young nuts. Their feeding damage is evident as scars on the surface of the older nuts. The damage results in reduction in size of nut and increases labour for dehusking the nuts. These mites being wingless have to depend on wind for dispersal and randomly land on the crown region of the palms. Following this they should be able to reach the area of the young nuts which is most suitable for them for feeding and breeding. Laboratory trials were carried out using 68 nuts of different age groups with varying parameters like distance between perianth and point of release of the mite, position of nut, colour and source of light. The success and failure of the mite in reaching the perianth was recorded. The success rate was found to be 88 per cent and failure rate was 12 per cent. The mite reached the perianth of nut successfully when the age of the nut was less than 150 days old. Significant positive relationship was noticed between the age of the nut and the time taken to reach the perianth ($r=0.707$; $p<0.001$; $n=37$).

However negative association was observed for the same if the age of the nut crossed 150 days. Significantly positive relationship was observed between distance of point of release of the mite to the perianth and time taken to reach the perianth ($r=0.370$; $p<0.003$; $n=60$). As the distance between the mite landing point and perianth increases, the time taken by mite to reach the perianth also increases. If the mite lands very near to the perianth, the mite directs its movement towards perianth and soon reaches the region beneath the perianth. The results indicated that the mite depends on some odour and cue emitted from the lower side of the perianth, hence it tracks the odour coming from perianth and moves towards perianth. The mite can move underneath perianth, irrespective of nut position *viz.,* perianth is above or below or is horizontal. The emission of odour and cue is maximum in younger nut and decreases as the nut grows older. The emission comes down to below threshold level in the nuts of over 150 days old, at which the mite are unable to detect and track the cue coming from the perianth. Therefore, the mite fails to reach peianth when nut is over 150 days old.

18

Account of An Architect Queen

MAULISHREE AGRAHARI
AND RAGHAVENDRA GADAGKAR

ABSTRACT

Queens of 4 post-emergence colonies of primitively eusocial wasp, *Ropalidia marginata* were observed for their propensity to acquire and handle foraged material on naturally occurring free foraging nest. The queens were inclined to acquire pulp loads than food loads. Queens fed larvae at low rates. The pulp loads were used for nest building. They build new cells more often than build walls of old existing cells. Queens did 59 of 64 events of "initiate new cell". This could indicate centralized control of nest building activity. In other primitively eusocial species like *Polistes,* the queens forage for pulp, build new cells, and lay eggs. *R. marginata* queens on the other hand do not forage but quite like queens of other primitively eusocial species, tend to acquire pulp loads from returning foragers. This observation suggests queens of post-emergence colonies of *R.marginata* might have adopted the strategy of acquiring loads from returning foragers rather than foraging for themselves. This could be a possible option which might prevent them from predation and other fatigue related risk. Comparing queen behaviour with other primitively eusocial wasp colonies seem to hold pointers towards our previous claim that *R. marginata* is relatively advanced by primitively eusocial standards.

19

Behavioural Difference and Diversity of Males and Females of Two Primitively Eusocial Wasps

RUCHIRA SEN AND RAGHAVENDRA GADAGKAR

ABSTRACT

In social Hymenoptera, the females carry out nest building, brood care and foraging, while males generally do not take part in any of these social behaviours and act solely as reproductives. The interesting phenomena of cooperation and altruism exhibited by the females have led to a plethora of theoretical and empirical studies. On the other hand, males in social Hymenoptera have been relatively neglected. Here we have studied two sympatric, primitively eusocial wasps, *Ropalidia marginata* and *R. cyathiformis,* and compared the behavioural profiles of males and females. Males of *R. marginata* stay on the nest for 3-12 days (6.0±1.9, n=89) and then leave once for all to lead a nomadic life, while *R. cyathiformis* males stay on their natal nests for their entire life span (16.7±5.6, n=11). During their stay on the nest the males of both species occasionally perform the following behaviours: solicit food, antennate nest, antennate another wasp, feed self, snatch things, fan wings, body jerk and wing jerk, dominance and subordinate behaviours. Using methodologies analogous to that used in ecological studies of species richness and diversity we show that females of both species are behaviourally more rich and more diverse than the males.

20

Activity Patterns of Grey Jungle fowl, *Gallus sonneratii* At Gudalur Range in Theni Forest Division, Western Ghats, Tamilnadu South India

RAMESH, N. AND M.C. SATHYANARAYANA

ABSTRACT

Activity patterns of Grey Jungle fowl, *Gallus sonneratii* were studied at Gudalur Range (9° 37′ N, 77° 16′ E) which is 23² km area in Theni Forest Division, Western Ghats, Tamilnadu from January to December 2005. A daily activity schedule was calculated from approximate 12 hours of day time observation. Activities such as foraging, vigilance, interaction, calling, body comfort (preening, wing flapping, leg stretching and tail shaking) display, incubation and resting were recorded in different age and sex classes of Grey Jungle fowl. Two types of feeding activities were recorded. The Grey Jungle fowls were observed feeding throughout the day. However, the bird was very active during early hours of the mornings and evenings and slower during mid day hours. The overall result revealed that the foraging activity ranged from 46.00% to 56.50% in sub adult female to adult female. The vigilance activity varied between 2.25% and 10.25% in juvenile and adult male while the resting ranged from 9.50% in adult female to 20.75% in juvenile. Minimum percentage of calling (0.50%) was observed in juvenile and maximum percentage of calling (4.50%) was observed in adult male. In general, the adult male produces more vocalization when compared to the other age and sex classes. However, adult female was mostly silent but occasionally gave alarm call. The body comfort of Grey Jungle fowl was minimum in adult female (14.75%) and maximum (18.75%) in juvenile. Greater display was observed in adult male (1.75%) and adult female (0.50%). Only adult female attended incubation duties (2.50%). The interaction activities ranged from 6.50% in adult male to 10.50% in sub adult female and juvenile. The activities varied between seasons viz., pre-monsoon, monsoon, post-monsoon and summer. Similarly, during breeding and non breeding seasons variation in different activities were observed.

21

Analysis of Route Learning Ability of Climbing Perch *Anabas testusineus*

AVINASH, T.P., V.V. BINOY AND K. JOHN THOMAS

ABSTRACT

Learning the route and spatial complexity is essential for the survival of all animals including fishes. The present study analyses the route learning ability and the influence of presence of landmarks as reference points on the route learning ability of a fresh water fish, the climbing perch (*Anabas testudineus*). Our results show that climbing perch learn the route provided very fast. They adapt very soon to the alterations in the learned route. Landmark cues have no influence on the route learning in this fish and they depend on the body-centered movement to reach the target.

22

Predator Avoidance Behaviour of Naive Juveniles of *Puntius Sarana*, A Freshwater Fish

JOHN SHIRON, V.V. BINOY AND K. JOHN THOMAS

ABSTRACT

We studied predator recognition and the cues involved in eliciting predator avoidance behaviour in induced bred, naive, juveniles of *Puntius sarana* (2 ± 1 cm; S.L.). The results show that the size of the stimulus fish is one of the major cues that elicit avoidance response in the juveniles. Naive juveniles of *Puntius sarana* avoided large predatory (*Anabas testudineus* S.L. 10 ± 1 cm) as well as non-predatory fishes (*Orechromis mossambicus* S.L. 10 ± 1 cm). However they failed to exhibit any type of avoidance response in presence of small (S.L. 5± 1 cm.) non-predatory fishes. When the size of the stimulus fish [gold fish, non predatory juveniles 2 ± 1 cm] was equal to that of the test fish, they preferred to remain near stimulus fish, instead of avoiding it.

23

Influence of Hunger on Food Stocking Behaviour of Climbing Perch, *Anabas testudineus*

BINOY, V.V. AND K. JOHN THOMAS

ABSTRACT

Fishes employ a variety of strategies to enhance their feeding efficiency. Full expression of many of these tactics is observed when the animal is hungry. The present study describes the food stocking behaviour of climbing perch, *Anabas testudineus,* a behaviour that has not yet been reported in any species of fish. The climbing perch stocks food in its mouth whenever it is available. Food deprivation for twenty-four hours increases food stocking, but not food intake. The second part of the study shows that the fish retrieved the pellets only when they are isolated on the same day of capture. Meanwhile the fish isolated after keeping in-groups (*Five in number*) for two months in the laboratory fails to exhibit this behaviour pattern. Influence of competition and the alteration of behaviour as a result of co-living in laboratory will be discussed.

24

Diel Activity and Diversity of Butterflies Visiting Nectar Host Plants in Dryland-scrub Ecosystem

JAYAPPA, J., M. MURALIKRISHNA, REVANNA REVANNAVAR , MAHADEVAPPA, K. DEVARAJ K. CHANDRASHEKARA AND A.R.V. KUMAR

ABSTRACT

Adult butterflies visit flowers, preferably with tubular corollas, to acquire nectar to meet their energetic needs. While several flowering plants offer nectar as rewards to insect pollinators, not all insect pollinators visit flowers indiscriminately. Thus each flowering plant has specific set of insect visitors determined by an interplay of various factors such as flower size, flower colour, corolla length, *etc.* This ensures that plants do not compete for the same set of insect pollinators. The insects associated with a flowering plant species either as pollinators or opportunistic visitors exploiting the flowers as nectar resource also avoid competition by visiting flowers at different times of the day, a temporal partitioning of resources. In this study we look at the butterfly community associated with plants inhabiting a dryland scrub forest ecosystem. Specifically we looked at two plants, *Duranta erecta* and *Clerodendrum* sp. during their respective flowering periods at the GKVK campus of the University of Agricultural

Sciences, Bangalore. The butterfly abundance, diversity and visiting frequencies were recorded on these plants from 0600 to 1900 h. Among the sixteen species, Blue Tiger, *Tirumala limniace* Cramer and *Euploea core* Cramer were the first to visit the flowers of the *Duranta erecta,* and were active even after dusk, till the end of the observations at 1900 h. Common banded Owl, *Hasora chromus* Cramer was the first species to visit the *Clerodendrum* flower followed by Common Rose and Tawny Coster, *Acraea violae.* The pattern of diversity of butterfly species at different hours of the day was found to be similar on both the plants. The diversity was low at 0600 h in early morning and increased steadily to reach the maximum at 1200 h on *Duranta erecta.* However, two peaks of diversity were observed on *Clerodendrum* at 1000 and at 1400 h. The species accumulation on both plants was computed and subjected to probit analysis to determine the time required for accumulation of 50 per cent of the species associated with the plant (SA_{50}). The time required for 50 per cent of the total species was different for the two plants with the SA_{50} being 08.23 and 09.65 h, respectively for *Duranta erecta* and *Clerodendrum.* The underlying causes for this difference could be the difference in the number of flowers, the size of flowers, the quantum of nectar and more importantly the pattern of nectar production through the day. The frequency of visits of each butterfly species was used to compute the 'activity breadth' using the Shannon Index. Three and four groups of butterfly species were observed on *Duranta erecta* and *Clerodendruma* plants following cluster analysis by Wards' method and these groups were significantly different. In conclusion, nectar host plants attract variety of butterfly species and their visit is not confined to single hour. It is important therefore to have different nectar host plants to increase the diversity of butterflies attracted in dryland-scrub forest ecosystem.

SECTION - 3
SOCIAL BEHAVIOUR & PHYSIOLGY OF BEHAVIOUR

25

Physiology of Behaviour of White Leghorn Chick (*Gallus domesticus*) Exposed to Pesticides

SOWMINI BALAKRISHNAN

ABSTRACT

Indiscriminate use of agricultural pesticides has created very serious health and environment problems in many developing countries. Industrial waste contaminates environment. The diversity of pollutants result in physiological effects at several levels including direct effects on breeding adults and developing embryos. Factors such as temperature, diet changes etc affect the physiology of behaviour in birds. The present study is undertaken to find out the effect of dimethoate on the behaviour and biochemical composition of vital organs of White leghorn chick (*Gallus domesticus*).

Behavioural alterations in the exposed chick are hyper activity, jerky movements, convulsions, loss of equilibrium and infiltration of blood. Present study indicated that the pesticide, even at sublethal concentrations may adversely affect behaviour and biochemistry of birds. Vital organs of control and treated chicks were analysed for the protein and carbohydrate content. Qualitative and quantitative variations were observed.

INTRODUCTION

Food production has been mainly dependent on the synthetic chemicals to combat the natural pests in agricultural areas. Indiscriminate use of agricultural pesticides has created very serious health and environmental problems in many developing countries. Industrial waste disposal contaminates environment. The diversity of pollutants results in physiological effects at several levels (Jalaluddin, 1987; Gregory *et al.,* 2005) including direct effects on breeding adults and developing embryos (Fry, 1995). The fungicide, Fenarimol after prenatal exposure altered several aspects of the reproductive system. The fenarimol placental transport increases from the 18th gestational day and is excreted via maternal milk (Castro *et al.,* 2006). Atmospheric emissions of Lead causes toxicity which is very expressive in embryos and young individuals (Rivero *et al.,* 2006). Previous studies indicate that only one exposure to lead in chick embryos, depending on the ingested dose can result in malformations of face and limbs, hydrocephaly (accumulation of fluid within the cranial cavity) and reduction in the eclosion (emergenc from larva to Pupa and Pupal stage to insects). Influence of parastic infestation and factors such as temperature (Krijgsveld *et al.,* 2003), diet change etc affect the physiology of behaviour in birds (Greg *et al.,* 2006).

Dimethoate, an organophophorous insecticide (water soluble) is used in poultry houses and in cattle rearing for control of a wide variety of sucking pests and lice infesting poultry. It is particularly effective against sucking insects such as aphids on trees and shrubs. It is also used as a residual wall spray in farm buildings for house flies. It is phytotoxic to certain varieties of sorghum, Chrysanthemum etc. This pesticide is available in aerosol spray, dust, emulsifiable concentration etc. It is formulated as 30% EC. Trade names include Rogodan, Rogor, Trimetion, Dimet, Fostion MM, Sevigor, Roxion. Dimethoate is moderately toxic by ingestion, inhalation and dermal absorption. It is readily absorbed through the skin and lungs. High environmental temperatures or exposure to visible UV light may enhance its toxicity. The organophosphate insecticides are cholinesterase inhibitors, highly toxic by all routes of exposure. It is rapidly metabolized in mammals but not in birds highly toxic to birds. In the environment and in sewage treatment plants it undergoes rapid degradation. It is a colourless, crystalline solid with a camphor like odor (mercapton). It will decompose rapidly when heated to temperatures above 80°C creating possibe explosion.

The present study is undertaken to find out the effect of Dimethoate, an organophosphorous pesticide on the behaviour and biochemical composition of vital organs of White leghorn chick *(Gallus domesticus).*

MATERIAL & METHODS

The alterations in biochemical and development parameters are among the measures of events most studied as possible biomarkers depending on the critical periods in the development (Castro *et al.,* 2006).

One day old White leghorn chicks *(Gallus domesticus)* were taken for study. They were obtained from Kattuppakkam near Chennai. A dozen of them were brought and divided into 2 groups each having six numbers. One group was treated as control and another group was taken for pesticide treatment. The control animals were kept healthy and both groups were maintained in cages.

Method of Study: Toxicity Studies - Oral administration of pesticides in chicks: Dimethoate was purchased from pesticide shops. One ml of pesticide was diluted with 100ml of distilled water (1%). 0.25ml of this diluted mixture was given to the chick. Chick died within two days. After that, 1ml of dimethoate pesticide was diluted with 500ml of distilled water (0.2%). 0.25ml of this diluted mixture when given to the chick, it survived for a week. But in 0.125ml concentration the chicks survived for more than 15 days. After two weeks of treatment the six chicks were sacrificed for quantitative analysis of protein in the liver and kidney.

Quantitative estimation of protein by Lowry's Method. (Lowry *et al.,* 1951): The principle of this method involves two steps. The carbonyl groups of protein molecules react with copper and potassium of the reagent to give a blue coloured copper potassium biuret complex. This complex together with tyrosine and phenolic compound present in the protein reduce the phosphomolydbate of the Folin reagent to intensify the colour of the solution (Lowry *et al.,* 1951).

RESULTS

There was no change in the weight or reduction of weight in the treated chicks.

Results are presented in Table 1 & Fig 1. Quantity of protein decreased in both liver and kidney of treated birds when compared to that of the control. The quantity of protein in the treated liver was 67.50 ± 0.2828 mg/gram and that of control was 85.00 ± 0.8185 mg/gram wet weight. Kidney protein levels were 70.00 ± 1.17 mg/gm wet weight in control chicks and the treated kidney tissue had 47.00 ±0.81 mg proteins/gram wet weight.

Behavioural alteration in the exposed chicks were hyper activity, jerky movements, convulsious, loss of equilbrium and infiltration of blood. This is in agreement with earlier workers who have reported the above symptoms and hemorrhages in the vertebral region of pesticide-exposed fish.

Table 1: Protein levels in the liver and kidney of one day old control and treated *Gallus domesticus*

S.No	Control Liver	Treated Liver	Control Kidney	Treated Kidney
1.	84.00	67.50	72.00	47.00
2.	85.00	67.65	69.90	48.00
3.	86.00	66.90	68.00	47.00
4.	86.00	67.70	69.10	46.00
5.	84.00	67.75	71.00	48.00
6.	85.00	67.50	70.00	46.00
Mean Value	85.00	67.50	70.00	47.00
Standard deviation	± 0.8185	± 0.2828	± 1.1704	± 0.8185

DISCUSSION

Protein Content in the Control and treated Tissues: The amount of protein declined vigorously in the tissues of experimental animals compared to control. The decrease in protein content in treated tissue of the chick may be due to their degradation to amino acids which are used in TCA cycle through amino transferase pathway to cope up with high energy demands during the state of depleted glycogen. Similar observations were made by Pushpa Raj *et al.,* (1993) in the Aldrin treated testis of *Rana hexadactyla*. According to Anjaneyalu & Sundaraj (1988), the Aflatoxin treated broilers exhibited a decrease in the levels of serum proteins. This may be either due to the activity of proteolytic enzymes or due to changed membrane permeability (Jalaluddin, 1987).

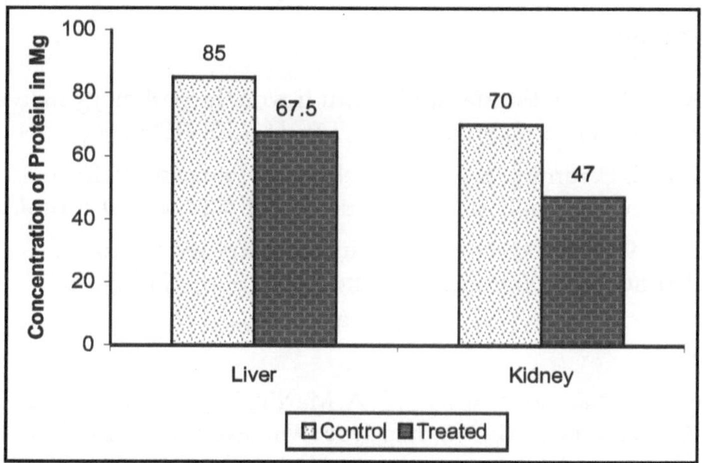

Fig. 1. Protein levels in the liver and kidney of control and pesticide treated domestic chicks

Previous studies demonstrated the massive impact of large populations of mites on the behaviour and physiology of laying hens (*Gallus domesticus*). Due to mites, the hens were losing blood. Hens compensated for it by increasing the production of erythrocytes. New erythrocytes were found to be larger and had a lower concentration of haemoglobin than mature erythrocytes.

Thyroid hormones T_3, T_4 and TSH were all affected by pesticide treatment (Gregory *et al.*, 2005). The effect of pesticides on thyroid function in 16-day old tree swallows (*Tachycineta bicolor*) and 12-day old eastern blue birds (*Sialia sialis*) was reflected in elevated plasma T_4 concentration and higher follicular epithelial cells relative to control chicks (Gregory *et al.*, 2005). It also indicated that plasma T_4 concentration was positively correlated with the total number of pesticide mixtures applied during egg incubation through chick rearing. In addition chicks were losing weight. Chromosomal alterations in bone marrow cells were also observed (Nehez *et al.*, 1983).

CONCLUSION

Present study showed that the pesticides even at sub lethal concentrations may adversely affect the biochemical nature of the tissues.

Much work remains to be done to elucidate human health impact of exposure to persistent pollution of organic pesticides.

References

Anjaneyalu. Y. & A. Sundaraj.1988. Studies on the pathology of liver in poultry. *Abstracts of Research on Poultry. TNAU VAS* – 1996.

Castro, V, S. Scramin & O. M. Mel. 2006. Fenarimol effects on the animal development and rat's reproduction. *J.Braz. Soc. Eco. toxicol.*1: 71-74.

Fry, D.M. 1995. Reproductive effects in birds exposed to pesticides and industrial chemicals. *Environ Health Perspect.*103 (Suppl 7): 165-171.

Greg, L., E. Green & C. Warwick. 2006. Abstract. *J. Appl.Anim.welfare Sci.* 9(1):. 41-58.

Gregory, J., C. Mayne, A. Bishop, P. A. Martin, H. J. Boer Mans & B. Hunter. 2005. Thyroid function in Nestling Tree swallows and Eastern blue birds exposed to non-persistant pesticides and PP-DDE in Apple orchard of southern Ontario, *Canada. J. Eco.toxicol.* 14(3).

Jalaluddin, S.D. 1987. Modulations in the biochemical nature of the body muscles of a fresh water teleost *Sarotherodan mossambicus* (Peter) due to "DET IOL" (A synthetic detergent). *Proc. Nat. con. Env. Impact on Biosystem.* January 1987. Loyola College, Madras.

Krijgsveld, K. L, H. Visserg. & S. Daan. 2003. Foraging behaviour and physiological changes in precocial quail chicks in response to low temperatures. *Physiol & Behaviour.* 79: 311-319.

Lowry,O. H., R. Brough, A. L. Farr & R. J. Randall. 1951. Protein measurement with folin phenol reagent. *Biol. Chem.* 193: 265-275.

Nehez, M, S., H. Schenfler & G. W. Fischer. 1983. Effect of Dimethoate and O-Demethyldemethoate on bone marrow cells of CFLP mice Regul. *Toxicol. Pharmacol.* 3(4) : 349-54.

Pushpa Raj, .V., T. Rajeswari & K. Jayantha Rao. 1993. Effect of sublethal concentration of Aldrin on the testis of *Rana hexadactyla. Bull. Env.Sci* 1X1: 10-13.

Rivero, L. B. D., J. C. Schatz, M. S. L. Carvalho, M. C. De Carvalho & Y. M. Rauh Muller. 2006. Lead acetate action in embryonic developmental pattern of *Gallus domesticus. J.Braz. Soc. Ecotoxicol* (1): 83-187.

26

Finding An Appropriate Index to Construct Dominance Hierarchies in the Primitively Eusocial Wasps, *Ropalidia marginata* and *Ropalidia cyathiformis*

ALOK BANG, SUJATA DESHPANDE, A. SUMANA AND RAGHAVENDRA GADAGKAR

ABSTRACT

Dominance hierarchy is an important component of several behavioural studies, especially studies pertaining to animals living in groups. There is a vast diversity in the methods of computation of dominance ranks, but very rarely the rationale behind using a particular method or index has been explained. In this study, we have looked at three dominance indices, viz. Index of Dominance (Premnath *et al.*, 1990), Clutton-Brock *et al.*,'s index (Clutton-Brock *et al.*, 1979) and David's Score (David, 1987). Apart from Index of Dominance that is formulated and used by our group to calculate dominance ranks in *Ropalidia marginata* and *Ropalidia cyathiformis*, which are primitively eusocial paper wasps, the other two indices were chosen based on popularity and recommendations by other groups working on dominance behaviour and hierarchies. Dominance ranks were obtained from the dominance-subordinate interactions in the colonies of these two species using all the three indices. The indices were examined based on two criteria: first, similarity between the ranks attributed by the three indices and second, number of unique ranks (i.e. ranks without ties). We found that all the indices attribute ranks that are similar to each other, with a significant positive correlation amongst them. However, in case of unique ranks, the Index of Dominance gives maximum number of unique ranks and significantly more than the other two indices, for both the species. Our results, thus, show that the choice of index to be used should be consciously decided depending on the properties of the index as well as the social biology of the species to which it is applied.

27

Scent Marking by Free-ranging Dogs (*Canis familiaris*)

PAL, S.K.

ABSTRACT

Two neighbouring dog groups that roamed freely throughout the day and night and whose home ranges as well as sexual experience were known, were selected to collect data on scent marking. Focal animals were followed 6 h per day. A total of 4680 h were devoted to collect data. The incidence of urine marking was greater than the incidence of simple elimination (c2 = 576.79, df = 1, P <0.05). The rate of urine marking was higher in adult dogs (c2 = 54.56, df = 1, P <0.05) as well as in adult males (c2 = 128.14, df = 3, P <0.05). The rate of marking was minimum in case of juvenile dogs, and it increased with the age. The seasonal mean (± S.E.) number of marking for individual dog was maximum (34.92 ± 6.30) in late monsoon and least in winter (19.77 ± 2.70). Thus, the seasons were significantly different from each other in relation to the frequency of urine marking (c2 = 33.07, df = 3, P <0.05). The incidence of urine marking was highest in the courting places, and the places were significantly different from each other in relation to the number of marking (F = 3.27; 4, 68; P <0.05). The males marked with higher frequency in the courting places showing the relation between marking and the reproductive status of the dogs. The incidence of marking mostly by the males near the territorial boundaries during trespassing by neighboring dog(s) showed evidence of territorial defense. Marking in the feeding places indicated the evidence for scavenging behaviour of dogs. To protect the pups, the females marked with a higher rate at the nest site. The dogs, especially the males, marked on strange objects to familiarize with the strange objects.

28

How Do Workers of the Primitively Eusocial Wasp, *Ropalidia marginata* Detect the Presence of Their Queens?

ANINDITA BHADRA, PRIYA L. IYER, A. SUMANA, SUJATA A. DESHPANDE, SAUBHIK GHOSH AND RAGHAVENDRA GADAGKAR

ABSTRACT

Primitively eusocial insect societies are typically headed by behaviorally dominant queens. However, in *Ropalidia marginata*, the queen is strikingly docile and behaviorally non-dominant. Interestingly, if the queen is removed from the colony, one of the workers (potential queen) becomes extremely aggressive within minutes, and eventually becomes the new queen, if the original queen does not return. The fact that PQs are not discernible in the presence of the queen, but become obvious within minutes of queen removal raises the intriguing question of how workers perceive their queen, or how she signals her presence to them. It is possible that the queen of *R. marginata* uses a pheromone, either volatile or non-volatile, to signal her presence to her workers. In a set of experiments, we separated a set of workers by a wire mesh partition from their queen and studied their response to such separation. It was obvious that the workers did not perceive the queen across the partition. So we concluded that the queen signal in *R. marginata* is non-volatile. We estimated the time taken for the

PQ to perceive the absence of the queen, and termed it the realization time and built a model incorporating the realization time, the decay time of the queen's signal and the average signal age. We checked if the realization times estimated through the model matched that obtained from real data. We found the rates of direct interactions to be too low; and so considered the possibility of signal transfer through relay. Using the Dijkstra's algorithm in simulations, we established the effectiveness of relay in such a system. Then we used data from 50 colonies to fit our model. We concluded that interactions, both direct and indirect, are not sufficient to transfer the queen's signal in *R. marginata*. Finally, we tested the possible role of the "rub abdomen" behaviour of the queen in transmitting her signal in the colony.

□□□

29

Transition of an Aggressive Potential Queen to A Docile Queen in the Primitively Eusocial Wasp *Ropalidia marginata*

ANIRUDDHA MITRA AND
RAGHAVENDRA GADAGKAR

ABSTRACT

Ropalidia marginata is a primitively eusocial paper wasp widely distributed in peninsular India. Colonies comprise of a queen who monopolizes reproduction, and several other females functioning as sterile workers. It is similar to other primitively eusocial species in the absence of morphological difference between queen and workers. Yet unlike other species, *R. marginata* queens are remarkably docile in behaviour, which is a characteristic of highly eusocial species. An interesting fact is that when such a docile queen is removed from the colony, one of the remaining individuals emerges as the potential queen, becomes very aggressive and goes on to become the next queen. This suggests that potential queens of *R. marginata* may use dominance to suppress worker reproduction, and later on use some other mechanism, possibly pheromones, and become established as queens once they develop their ovaries. Thus, this system provides a unique model to study the shift in control of worker reproduction from physical inhibition by queen to pheromonal queen signal. We hypothesise therefore, that there should be a negative correlation between dominance behaviour and ovarian development of potential queens. In the present study we provide evidence in support of this hypothesis. In future studies, we propose to also measure the development of pheromonal signal by potential queens. Such studies, in addition to giving evidence for transition in the nature of regulation of worker reproduction from physical control to pheromonal signalling, will also give clues regarding evolution of primitively eusocial to highly eusocial state in general.

30

An Observational Study on the Behaviour of Chimpanzee (*Pan troglodytes*) in Captivity At National Zoolgical Park, Delhi Zoo

SAMINA HISAMUDDIN FAROOQI
SATISH KUMAR AND H.S.A. YAHYA

ABSTRACT

A pair of chimpanzees (*Pan troglodytes*), consisting of one adult male and an adult female was systematically observed in captivity over 204 hours. Data were collected using Focal Animal Sampling Technique (Altman, 1974) and compared with the information available in wild conditions. In addition to their sex-specific differences, both the male and female chimpanzee differed from each other in their behavioural repertoires, such as the male, Rustam spent maximum time (50.6%) in locomotion, whereas, the female chimpanzee, Ruby spent in resting (61.3%). Most behavioural repertoires were remarkably similar, both qualitatively and quantitatively, to that of wild chimpanzees.

31

Stress-induced Behavioural Changes in A Novelty Test in the Domestic Chick

RESHMA SHETTY AND MONIKA SADANANDA

ABSTARCT

The effect of inescapable foot shock on different behaviours was examined in two- and five-day old chicks, *Gallus domesticus* in the open-field. The apparatus consisted of a 90cm diameter arena placed in a darkroom with targeted illumination. Recordings were made over a period of 10 minutes either mechanically or manually in a semi-automated manner. The inescapable foot shock was administered intermittently at an intensity of 30-40mA for 20 milliseconds over a period of 30 minutes. Chicks were stressed once on 2phd (post-hatch day) or on 4 successive days until 5phd. Each stressed group had same-aged controls. During exposure to a novel open-field environment, which is a stressor and indicator of fear levels in birds, the behavioural measures screened for, included latency to first step, locomotor activity, call and peck frequencies and field movement. The data were tested for significant differences using 2-way ANOVA or non-parametric tests followed by post-hoc tests. Behavioural measures such as call frequency and movement were positively related with age, while peck frequency and latency to first step were negatively correlated. When compared to same-aged controls, stressed birds demonstrated negative behavioural patterns in locomotion (number of steps per minute), peck and call frequencies, but were positively related to latency to first step, and freezing. In field movement, stressed birds exhibited thigmotaxis, that is, the distance covered is higher in outer fields than in inner fields, while normal birds either show no particular preference for the exterior or the interior of the arena. Results from this study indicate that short, inescapable foot shock over a short or long time period leads to enhanced fearfulness in open-field in birds.

32

A Study on the Reproductive Physiology of A Captive Pair of Chimpanzee (*Pan troglodytes*) in National Zoological Park New Delhi

SAMINA H. FAROOQI, H.S.A. YAHYA AND IQBAL PERWEZ

ABSTRACT

Disturbed breeding and mating is quite common in captive chimpanzees (*Pan troglodytes*). To establish a successful breeding program in captivity for this species, assessment of reproductive status is indispensable. The present study was carried out on a pair of common chimpanzees in National Zoological Park, New Delhi during 2002-04 with the objectives that some hormonal disorder could be the possible cause for the breeding failure of the chimps in captivity and to ascertain the effect of this hormonal problem on the sexual behaviour. Data were collected and blood samples were taken to analyze the hormonal profile. Since there was not enough data on the hormonal profile of wild or captive chimpanzees, we compared our results with that of human profile because the hormonal profiles in chimpanzees are known to be very similar to those of humans. We obtained the values that correspond to the normal levels in human serum. The paper details our findings and offers recommendations for conservation of captive chimpanzees.

33

Expression of Red/Green Color Sensitive Opsin in Chiroptera A Prime Report

KANAGARAJ, C AND K. EMMANUVEL RAJAN

ABSTRACT

Bats display biological rhythm associated with flight activity but the photoreceptors that mediate this behavioural pattern are largely unknown. In general, vertebrate retina contains two types of photoreceptors: rods, that function at low light levels and mediate scotopic vision, and cones, that function in day light and mediate photopic vision and thus provide the basis for colour vision. Chiropterans are adapted to nocturnal life style; their retinas are heavily dominated with rod cells and significant number of cone cell populations. Behavioural observation shows that bats are sensitive to UV-light, which was later supported by sequence data from short wave length opsin and L/M opsin. To demonstrate the expression of opsin, we tested retinal and brain tissue extracts against the anti-red/green opsin antibody. The antibody recognized a protein with an apparent molecular mass of 36 kDa in both tissues. This is the first report of red/green like opsin expression in bat brain. Taken together, these results suggest that a red/green color vision may exist in both Megachiroptera and Microchiroptera, and may play a vital role in visual task and circadian rhythm. Further studies are needed to understand the spectral sensitivities of retinal opsin and the regulation of extraocular opsin with other photo entrainment genes.

SECTION - 4
SOCIAL BEHAVIOUR & PHYSIOLGY OF BEHAVIOUR

34

Feeding Behaviour of Cardamom Shoot and Fruit Borer (*Conogethes punctiferalis* Guanee) Larvae in Laboratory

THYAGARAJ, N.E. P.K. SINGH AND
A.K.CHAKRAVARTHY

ABSTRACT

Laboratory studies were conducted on feeding behaviour of cardamom shoot and fruit borer (*Conogethes punctiferalis* Guanee) larvae during 2002 at Mudigere, Karnataka. *Conogethes* larvae cause injury to the pseudo stem/ stalk, panicle and capsule of cardamom. Eight alternate host plant species were offered for their biting and feeding separately and also together. First and second instar larvae showed biting response on *Alpinia,* but discontinued after the second instar. Third and fourth instar larvae fed and continued feeding on cardamom and *Hedycium* spp. When fifth instar larvae were tested on stalks of eight plant species, larvae fed on cardamom and *Alpinia* spp. and continued to feed on both the plants. When the stalks of all the eight plant species were offered together, all the larval instars fed and continued feeding on cardamom and *Alpinia* spp. only. From this study it is clear that there is a difference in feeding preference among larval instars.

INTRODUCTION

Cardamom (*Elettaria cardamomum* Maton) is an important spice crop, popularly known as "Queen of spices ". This crop is a major foreign exchange earner for India and finds its use in many ways (Thyagaraj, 1989). Over 72 insect pests have been recorded on cardamom so far (Siddappaji & Reddy, 1972; Chakravarthy & Khan, 1987). Out of 72 insect pests, only two are considered as economically important i.e., shoot and fruit borer and thrips. The detailed bio-ecology and management aspects of these two pests have been studied (Thyagaraj, 2002). Between the two pests, shoot and fruit borer causes damage to cardamom shoot, panicle and fruit and has got wide range of host plants for its survival in ecosystem. Keeping this in view, present study was conducted to know the feeding behaviour of its larval stages on different host plant species.

MATERIAL & METHODS

A plastic container (15 x 10 cm) with filter paper at bottom and having a perforated lid was used to study the feeding behaviour. Separate container for each test material was maintained. In each such container one larva of each instar was released using No.1 wet camel hair brush and ten such boxes were maintained per host plant part. In a set of containers, all the eight plant stalks were maintained and all instars were released. One larva was released into each container and maintained. Constant observations were recorded for their biting and feeding response and behaviour under laboratory conditions. Larvae were selected from the mother culture maintained to study the bio ecology of the pest in the laboratory. In one set of experiment, all stages of larvae were made to starve for 6-10 hrs and then released on the host plant part and observed for their biting and feeding behaviour.

RESULTS & DISCUSSION

The experimental results clearly indicated that the *Conogethes* larvae exhibit different types of feeding behaviour on different host plant species. First and second instar larvae showed biting response on *Alpinia*, but discontinued after the second instar (Table 1). Third and fourth instar larvae fed and continued feeding on cardamom and *Hedycium* spp. Out of eight plant species tested, *Hedycium* and *Alpinia* species were preferred (Table 2). When fifth instar larvae were tested on stalks of eight plant parts, they continued to feed on *Hedycium* and *Alpinia* species (Table 3). Wang & Cai (1997) reported feeding habit of *C. punctiferalis* on different crops in different places with which the present findings are in aggreement. When all the instar larvae were released on all the plant species simultaneously, larvae showed a greater response of biting and feeding on the cardamom plant (stalk) compared to other plants (Tables 1,2 & 3) (Thyagaraj, 2002).

Table 1. Orientational and biting behaviour of first and second instar *Conogethes* larvae on stalks of eight plants

Plant/Part	% Larvae orientating and palpating *** First instar **	Second instar larvae
Elettaria cardamomum	18.00 [a]	9.00 [b]
Curcuma longa	12.00 [b]	0.00[b]
Gingiber officianale	1.00 [b]	0.00 [b]
Ricinus communis	6.00 [b]	6.00 [b]
Amomum muricatum	7.00 [b]	0.00[b]
Colacasia sp.	6.00 [b]	1.00[b]
Alpinia neutans	15.00[b]	0.00[b]
Hedycium sp.	8.00[b]	2.00[b]

- Mean of four replications; 8 larvae per replicate;

** Length of Stalk = 15 mm; Figures followed by the same letters are not significant by X^2 test at 1 d. f,

*** = not all larvae orientated and palpated, some were in between.

Table 2: Biting and feeding of third and fourth instars *Conogethes* * larvae on stalks ** of eight plant species.

Plant/Part	% Larval biting on the stalk (Palpating) ***	
	Third instar	Fourth instar
Elettaria cardamomum	20.00[a]	12.00[a]
Curcuma longa	3.00[b]	3.00[b]
Gingiber officianale	8.00[b]	5.00[b]
Ricinus communis	4.00[b]	10.00[b]
Amomum muricatum	2.00[b]	4.00[b]
Colacasia sp.	3.00[b]	5.00[b]
Alpinia neutans	4.00[b]	4.00[b]
Hedycium sp.	18.00[a]	9.00[b]

- Mean of four replications; 8 larvae per replicate;

** Length of Stalk = 15 mm; Figures followed by the same letters are not significant by X^2 test at 1 d. f,

*** = not all larvae orientated and palpated, some were in between.

Table 3. Initial biting and continued feeding responses of fifth instar *Conogethes* * on stalks ** of eight plants.

Plant/Part	% larvae *** responding	
	Initial biting	Continued feeding
Elettaria Cardamomum	19.00[a]	19.00[a]
Curcuma longa	5.00[b]	4.00[b]
Gingiber officianale	9.00[b]	7.00[b]
Ricinus communis	10.00[b]	9.00[b]
Amomum muricatum	3.00[b]	5.00[b]
Colacasia sp.	11.00[b]	14.00[a]
Alpinia neutans	13.00[a]	13.00[a]
Hedycium sp.	10.00[b]	8.00[b]

• Mean of four replications; 8 larvae per replicate;

** Length of Stalk = 15 mm; Figures followed by the same letters are not significant by X^2 test at 1 d. f,

*** = not all larvae orientated and palpated, some were in between.

CONCLUSION

Conogethes larvae have definite feeding pattern and preference; *Alpinia* and *Hedycium* spp. were the two plant species preferred. Cardamom is most preferred food compared to other alternate host plant species. Although, they showed some responses of biting and feeding on alternate host plants initially they discontinued later. Even under starved conditions similar trend of feeding was observed. Out of eight plant species tested, larval instars showed feeding differences and cardamom was the most preferred at all the instars.

References

Chakravarthy, A. K. & M. M. Khan. 1987. Innovative tools for protecting cardamom from herbivores: Plea for crop management. *Quarterly News letter Asia and pacific plant Protection commission.* 30 (34): 12-17.

Siddappaji, C. & Reddy. 1972. Some new insect pests of cardamom and their control. *III rd International symposium on sub tropical Horticulture.* To day and Tomorrow's Printers and Publishers, New Delhi. PP. 72-73.

Thyagaraj, N. E. 1989. *Bio-ecology and Management of cardamom root grub (Basilepta fulvicornae Jocaby) in Karnataka.* M.Sc. (Agri.) Thesis., University of Agricultural Sciences, Bangalore. 110 pp.

Thyagaraj, N.E. 2002. *Integrated Management of some important Cardamom (Elettaria cardamomum M.) pests in hill region of Karnataka, South India*. Ph.D Thesis, Dr.B.R.Ambedkar University, Agra. 226 pp.

Wang, Y. X. & R. T. Cai. 1997. The emergence and control of *Conogethes punctiferalis* for You nai plum variety. *South China Fruits* 26 (3): 45.

35

Species Diversity of Rodents in Bamboo Flowering Areas of Mizoram and their Feeding Behaviour

AZAD THAKUR, N.S., A. N. SHYLESHA
AND D. KUMAR

ABSTRACT

Survey conducted in different bamboo flowering areas of Mizoram and Meghalaya revealed that rats collected from Lunglei and Saiha district of Mizoram were identified as albino rat and *Rattus sikkimensis*. Rodent activities in bamboo flowering areas were observed from April to September, thereafter, their activities started declining till March. Analysis of Bamboo fruits (*Melocanna baccifera*) indicated that two essential amino acids viz., tryptophan (205 mg/ g N) and methionin (24.94 mg/ g N) were prominent and N content in 100 mg sample was 1.442%. Feeding trial conducted with bamboo fruits and other materials with additives (mustard oil and pure ghee) showed that, rodents preferred (per 100 g body weight/day), maize > rice > gram with or without mustard oil to bamboo fruits, which reflected that the bamboo fruits were the least preferred. It was also observed that feeds mixed with pure ghee were the least preferred foods. No preference was observed for feeding on bamboo fruits compared to the conventional feed.

INTRODUCTION

Bamboo occurs in great abundance and variety in South and South east Asian countries. India has the distinction of being the second highest bamboo producing country after China and more than 55% of the crop is produced in North east India (NEH). Bamboo spreads over an area of 10.03 million hactres, which constitutes 12.8% of the total forest cover of India. The total bamboo cover in NEH region has been reported to be 38,197 km^2. Among seven sister states of NEH region, Mizoram occupies largest forest area under bamboo (30.8%), followed by Tripura (27.13%) and Meghalaya (26.0%). There are 16 genera and 58 species of bamboo in the region and about 26 different types of bamboo are reported to grow in Mizoram. *Melocanna baccifera* dominates over other species and constitutes approximately 95% of the total growing stock in Mizoram. In the region, bamboo plays an important role in rural and socio-economic development of the indigenous communities. Tribal communities of this region use this potential resource for food, shelter, furniture, handicraft, medicines and for various ethno-religious purposes.

The flowering of bamboos in these areas is cyclical with bamboo dying after seedfall. Rodent populations multiply tremendously in the presence of abundant bamboo seeds. Once these are exhausted they migrate to agricultural fields, devastating crops and causing food shortage. This year a spectre of gloom is haunting the North east region once again as the phenomenon of bamboo flowering is occurring in different parts of the North-east. Gregarious flowering of *Muli* bamboo is spread over an area of 18,000 sq km in India's north-eastern states of Mizoram (the epicentre), Tripura, Manipur, Arunachal Pradesh, Meghalaya and parts of Assam.

Bamboo flowering : The mechanism of the timing of bamboo flowering is not yet fully understood. Some bamboos flower only frequently, others rarely and some never flower in human memory. Depending upon the bamboo species, the flowers appear at intervals of several years, if not decades or centuries. It is one of nature's baffling mysteries. Since bamboo is anemophilous and wind pollinated it has many flowers at anthesis at the same time for successful spread of the pollen.

Two types of famines following periodical bamboo flowerings viz., *Mautam* and *Thingtam* are reported from Mizoram. Mautam is associated with *"Mao"* (*Melocanna baccifera*) species of bamboo (Fig.1), which flowers in 48 ± 2 years cycle and *Thingtam* is associated with flowering of *"Rawnal"(Dendrocalamus longispathus)* and *"Rawthing" (Bambusa tulda)* species. The gap between *"Mautam"* to *"Thingtam"* is 18± 2 years and *"Thingtam"* to *"Mautam"* is 28-30 years approximately (Table 1).

Table 1. Periodic bamboo flowering in Mizoram

Bamboo species	Period	Type of famine
Bambusa tulda / Dendrocalamus longispathus	1880-1884	*Thingtam*
Melocanna baccifera	1910-1912	*Mautam*
Bambusa tulda / Dendrocalamus longispathus	1924-1929	*Thingtam*
Melocanna baccifera	1958-1959	*Mautam*
Bambusa tulda / Dendrocalamus longispathus	1976-1977	*Thingtam*
Melocanna baccifera	2005-2007	*Mautam*

As observed, bamboo flowering starts in October-December, immediately after the rainy season (September-October) recedes. Initially there are many young inflorescences and within a few weeks whole clumps get transformed into huge inflorescences. There is wind pollination and seed-set starts in January-February and seed shed starts by the end of March. In the following months (March-May) during summer, there is an increase in seed-shed. By the end of summer abundant seeds are seen on the forest floor. Bamboo seeds germinate in a few weeks after rains start and the layer of bamboo seeds on the forest floor is replaced by a dense green carpet of bamboo seedlings.

Fig. 1. *Melocana baccifera*

Fig. 2. *Dendrocalamus hamiltonii*

IMPLICATIONS OF BAMBOO FLOWERING

People in North east India and elsewhere in the world, believe that bamboo flowering is indicative of famine. Such famines are common in some East Asian countries like Myanmar and Japan, and also southern Africa. In the Indo-Myanmar frontier tract, especially in Mizoram, they have caused much devastation. When bamboo flowers, famine, death and destruction will soon follow, goes a traditional saying in Mizoram. From time immemorial many beliefs are associated with bamboo flowering. In general, bamboo flowering is considered as a bad omen. Reference to this can be found in the *Mahabharata* written more than 5000 years ago. In 1959, bamboo flowering set off a chain of socio-political events in the state that ultimately led to insurgencies against the Indian Union which spanned over two decades, the reason being inadequate government help to combat the famine that followed bamboo flowering. Generally the following implications of bamboo flowering could be forecasted:

Effect On Environment

Ecological havoc: This strange phenomenon causes ecological havoc because of two reasons. First, bamboos die after flowering and it will take at least several years before bamboo plants re-seed, leaving bare exposed soil which could be disastrous in mountainous States and also lead to food scarcity. Second factor is that rats feed on the flowers and seeds of dying bamboo trees, which is said to activate a rapid birth rate among rodents, which in turn leads to a huge rat population feeding on agricultural crops in fields and food stored in granaries resulting in food shortages, starvation and death. The phenomenon of fluctuations in mammalian population, especially in rodents consequent to bamboo flowering has been recognized all over the world.

Rodent outbreak: The popular belief is that the gregarious flowering of bamboo produces large quantities of seeds, resulting in population explosion of rats (having short life cycles), which in turn leads to famine. People of Mizoram recognized two types of famines viz., *Mautam* and *Thingtam* of which, *Mautam*" is said to be much more severe. All the bamboos flower at the same time and the rat population increases suddenly over extensive areas causing famine. First record of "*Mautam*" was in 1864 when rats multiplied spontaneously and hundreds of Lushais died as a result of famine. A similar outbreak of rats associated with bamboo flowering was reported in 1910-1912. The flowering of *B. tulda* and *D. longispathus* during 1924-1928 stretched over a period of 4-5 years causing "*Thingtam*" with gradual buildup of rat population inducing more of scarcity than famine.

Bamboo flowering in 1958-59 resulted in an outbreak of rodents which spread over an area of 1000 square miles and developed into a "Plague" in Mizoram. The huge number of rats ate up food stock and standing crop. The rodent outbreak spread further to neighbouring area of Cachar (Assam) and Tripura where about 1000 acres of green fields were devastated resulting in famine. The rats devoured paddy, chilli, betel nut, cotton, chest nut, fruit, millet, bamboo shoots, tobacco stem, ginger and certain grasses too. Simultaneously all the crops were eaten up resulting in famine.

Effect on Rural Livlihood

A thriving economy revolves around bamboo. Presently bamboos constitute an important industrial raw material and are vital to the economy of tribals. The pulp and paper industry, construction, cottage industry and handloom, food, fuel, fodder and medicines annually consume around 22 million tonnes of bamboo. The present mass bamboo flowering will eliminate the entire standing stock of culms within 2-3 years. Processing industries, local weavers and tribals will lose their source of raw material. If this resource is left un-harvested and go waste by dying, the resulting loss shall be around Rs. 12,000 million.

RODENT SPECIES DIVERSITY IN BAMBOO FLOWERING AREAS OF MIZORAM

Survey conducted in different areas of Mizoram and Meghalaya showed that there was sporadic to mass flowering of *Melocanna baccifera* in both the states but no upsurge of rodents was noticed. During survey, rats collected from bamboo flowering areas were identified Zoological Survey of India, Kolkota (Table 2). It was observed that rodent activities increased in paddy fields/*jhum* fields during its ripening stages i.e. October and November, which coincided with bamboo flowering.

Comparative nutritive values of bamboo fruits and paddy : As paddy is common food of rats in NE Region, both paddy and bamboo fruits were evaluated in the laboratory for their nutritive values. Analysis of nutrient contents of *M. baccifera* fruits from Meghalaya and Mizoram showed that out of 20 amino acids only tryptophan (205 mg/g N) and methionin (24.94 mg/g N) were present which were not sufficient to induce enhanced reproductive behaviour. Carbohydrate content and dry matter were quite low in bamboo fruits as compared to paddy. Whereas, lipid content was higher in bamboo fruits but proteins and mineral contents were on par with paddy. (Table.3).

Table. 2. Rodent species diversity in bamboo flowering areas of Mizoram

Common name	Scientific name	Locality
Mackenzie's rat	*Rattus mackenziei* Thomas	Kolasib
House rat	*Rattus rattus* Linnaeus	Kolasib and South Hlimen
Himalayan rat	*Rattus nitidus* Hodgson	South Hlimen and Kolasib
Himalayan Chest nut rat	*Niniventer fulvescens* Gray	South Hlimen
Lesser Bandicoot rat	*Bandicota bengalensis* Gray	Bukvannei
Sikkim rat	*Rattus sikkimensis*	Saiha
Albino type rat	*Rattus norvegicus*	Lunglei

Table 3. Nutrients of bamboo fruits and paddy

Sample	Moisture %	Dry matter %	Ash %	Protein %	Carbohydrate %	Lipid %
Melocanna baccifera	72.84	27.16	3.68	9.01	26.90	7.80
Paddy	10.26	89.74	3.31	9.84	74.52	1.44

The fruits of *M. baccifera* (Fig.3) were comparable to paddy in their protein content. It was observed that when fruits of *M. bacifera* and paddy (as check) were fed to *Rattus nitidus,* it was recorded that bamboo fruits had no obvious effect on breeding of experimental rats.

Fig. 3. Bamboo fruit

Food preference of Rattus : The consumption of bamboo fruits of *M. baccifera* by rodents was believed to increase the fertility of rodents resulting in increased population densities. Feeding trials conducted under laboratory conditions with bamboo fruits, other feeds and other feed with additives (mustard oil and pure ghee) showed that, rodents preferred (per 100 g body weight/day) maize (9.72 g) > rice (9.42g) > gram (8.64g) with mustard oil and maize > (8.25g) rice > (8.14 g) > gram (7.59g) without additives to bamboo fruit (1.73 g/100g body weight/day), which reflected that the bamboo fruit was the least preferred food material (Table 4). No preference was observed for feeding on bamboo fruits compared to the conventional feed. Feeding behaviour and nutritive values showed that the bamboo fruit did not have any effect on the reproductive behaviour.

Table 4. Comparative feeding preference of *Rattus* spp.

Food	Initial rat wt (g)	Consumption per 100 g body wt/day	% gain/reduction in wt (after 60 days)
Maize	164.10	8.25	+ 14.79
Maize + Mustard oil	174.40	9.72	+ 16.14
Maize + ghee	180.40	6.30	- 6.21
Rice	159.33	8.14	+ 12.27
Rice + mustard oil	186.83	9.42	+ 13.32
Rice + ghee	186.90	6.14	- 5.08
Gram	137.33	7.59	+ 3.31
Gram + mustard oil	157.98	8.64	+ 4.23
Gram + ghee	188.67	5.07	- 4.97
Bamboo fruit	143.67	1.73	- 15.94*
Paddy	194.65	6.50	+ 10.21

* Rats died after 4 days

Feeding of maize with mustard oil resulted in the highest per cent body weight gain (16.14%) in a period of 60 days, which was followed by maize (14.79%) > rice with mustard oil (13.32%) > rice (12.27%) > gram with mustard oil (4.23%) > gram (3.31%) > paddy (10.21%). On the other hand, feeding of bamboo fruit resulted in 15.94% reduction of body weight which died in four days. Reduction in body weight of experimental rats was also recorded in case of maize with ghee additive (6.21%), rice with ghee (5.08%) and gram with ghee (4.23%).

CONCLUSION

It is therefore, concluded that there is no definite relation between bamboo fruits and rodent breeding because, bamboo fruits do not contain higher nutritive value than paddy. Moreover, these fruits do not contain estrogenic substances in required quantity to increase breeding as seen in laboratory trials.

36

Developmental Behaviour of Rice Hispa, *Dicladispa armigera* Oliver (Coleoptera: Chrysomelidae)

VIJAY KUMAR, L., S. U. PATIL AND A. K. CHAKRAVARTHY

ABSTRACT

An investigation on rice hispa was made during 2005 and 2006 at Agricultural Research Station, Kanakanady, Mangalore. The incubation period varied from 3 to 5 with a mean of 4.78 ± 0.77 days. The duration of total larval period varied from 6.95 to 15.75 with a mean of 13.06 ± 3.42 days. The pupation took place in the larval mines within the leaf. The pupal period varied from 4.08 to 6.25 with a mean of 5.23 ± 0.86 days. The adults of *Dioladispa armigera* emerged from pupal abode within the leaf by making a slit on the leaf surface. The number of matings in the beetles varied between 3 to 7 depending upon the longevity of the partner. The females lived longer than males. The adult longevity of males varied from 21.50 to 37.50 with a mean of 36.84 ± 6.85 days. Similarly the longevity of females varied from 22.80 to 54.36 with a mean of 46.63 ± 7.02 days.

INTRODUCTION

Dicladispa armigera Oliver (Colepotera: Chrysomelidae) is one of the important pests of rice in India. It is a major pest of rice in Assam,

Andhra Pradesh, Bihar, Madhya Pradesh and West Bengal and is a minor one in Kashmir, Himachal Pradesh, Maharashtra, Orissa, Uttar Pradesh and Karnataka (Thakur *et al.*, 1979; Das, 1980). The outbreak of the pest in 1962 caused a severe damage in different parts of North eastern states and 6,437 ha were affected by this pest in Sambalpur, Orissa alone (Rao *et al.*, 1971). In recent times the pest has been reported almost in all rice growing areas of southern India. In Karnataka, for the past one decade it is a major problem particularly during wet season. Considering the economic importance of this pest the present study was undertaken to know the susceptible stages in its life cycle to develop economically feasible management strategies.

MATERIAL & METHODS

An investigation was made during 2005 and 2006 at Agricultural Research Station, Kanakanady, Mangalore ($12^0 54'$ N latitude, $74^0 51'$ E longitude and 30 m AMSL). The adults were collected from the field and maintained on susceptible rice genotype, *Jaya* under caged conditions. In each cement pot ($0.2m^3$) 4 hills (8 seedlings) were maintained. Freshly emerged pairs were allowed to mate under greenhouse conditions in a aluminum clip cage (2.5 x 2.0 cm) with plastic lid at one end and muslin cloth with rubber band on the other side. The proximal end above the lid was placed on the leaf lamina of 5-8 leaves. The duration of various developmental stages and adult longevity were recorded by using this cage. To study the larval stages, trapping device made from cylindrical iron frame with an outer covering of perforated polythene bag having loop fasteners at end was slipped around 3 to 5 leaves. The leaves were examined every morning and developmental stages of the insects were recorded in aluminum foils tagged to the leaves. The observations were recorded at six-hour intervals and the data tabulated.

RESULTS & DISCUSSION

The data obtained during two consecutive years on the biological parameters of rice hispa were pooled and are presented in Table 1.

Eggs: The eggs were laid within the rice leaf epidermal layer, usually on the dorsal surface, in small numbers (20-45). The number of eggs laid daily varied from 3 to 20 per female. The total number laid during a period of 10.28 to 13.54 days varied from 43 to 102 with a mean of

78.00 ± 6.85 eggs per female. The oviposition commenced 2 to 3 days after emergence (3.86 ± 0.48 days) and continued for 11 to 20 days. The oviposition period varied from 10 to 14 days with a mean of 12.36 ± 1.21 days. Over 50 percent of the eggs, often about 90 percent, were deposited within the first 7 to 8 days consecutively from the date of commencement of egg laying, and thereafter the rest, at intervals of 2 to 4 days. The female lived for another 9 to 21 days with a mean of 16.26 ± 1.86 days. The upper half of the leaf was usually preferred by the female for egg-laying, and seldom the lower half. The incubation period varied from 3 to 5 days with a mean of 4.78 ± 0.77 days.

Larva: The larvae passed through four instars and moulted thrice as confirmed from observations of exuviae of the same set of larvae inside the leaf tunnel. The larval movement inside the leaf may be easily seen by placing it against light. The first instar larvae moulted after 1 to 3 days with mean of 2.46 ± 0.90 days. The second instar similarly moulted after another 1.20 to 3.50 days with a mean of 2.60 ± 0.86 days. The third instar usually took 1.75 to 4.25 days with a mean of 3.75 ± 0.97 days. The fourth or last instar larvae took 3 to 5 days with a mean of 4.25 ± 0.69 days. The duration of total larval period varied from 6.95 to 15.75 days with a mean of 13.06 ± 3.42 days. The mortality among the larvae was very high and only 10-15 percent lived to pupation stage (Table 1).

Pupa: The pupal stage of *D. armigera* lasted 4.08 to 6.25 days with a mean of 5.23 ± 0.86 days. Although morphological differences in separating the sexes in the pupal stage were not attempted, the males looked smaller than females. The pupation took place in the larval mines within the leaf (Table 1).

Adult: The adults of *D. armigera* emerged from pupal abode within the leaf by making a slit on the leaf surface. In the greenhouse, the emergence started with males preceding the females, sometimes day or two earlier and for the first 3 to 4 days a preponderance of males was observed. Observations on emergence revealed that during the second fortnight of August, the most favorable time appeared to be the hours between 6.00 A.M and 8.00 A.M.

Mating in captivity commenced 1 to 2 days after the pair was enclosed in the cage. The number of matings in the beetles varied between 3 to 7 depending upon the longevity of the partner. As regards to longevity, the females were found to live longer than males. The adult longevity of males varied from 21.50 to 37.50 days with a mean of 36.84 ± 6.85 days. Similarly the longevity of females varied from 22.80 to 54.36 with a mean of 46.63 ± 7.02 days (Table 1). Tiwari & Dubey (2001) also observed the commencement of mating in hispa after 3 to 4 days of emergence. The number of mating and duration were higher in the afternoon and morning as compared to evening (Deka & Hazarika, 1996, 1998).

Table 1. Biological parameters of rice hispa

Life stages	**Duration (in days)**	
	Range	**Mean ± SD**
***Incubation period**	3.56 - 5.85	4.78 ± 0.77
****Larval development**		
I instar	1.00 - 3.00	2.46 ± 0.90
II instar	1.20 - 3.50	2.60 ± 0.86
III instar	1.75 - 4.25	3.75 ± 0.97
IV instar	3.00 - 5.00	4.25 ± 0.69
Total larval development	6.95 - 15.75	13.06 ± 3.42
Pupal period	4.08 - 6.25	5.23 ± 0.86
Adult		
**Pre-oviposition	2.75 - 4.05	3.86 ± 0.48
**Oviposition	10.28 - 13.54	12.36 ± 1.21
Longevity		
Male	21.50 - 37.80	36.84 ± 6.85
Female	22.80 - 54.36	46.63 ± 7.02
Fecundity	43.00 - 102.0	78.00 ± 6.85
Total life cycle	14.59 - 27.85	23.07 ± 4.05

* Mean of 40 observations **Mean of 30 observations

References

Das, B. B. 1980. A note on the hispa epidemic in the field of Tripura. *Indian J. Entomol.* 42: 530-531.

Deka, M. & L. K. Hazarika. 1996. Mating behaviour of *Dicladispa armigera* Oliver (Coleoptera: Chrysomelidae). *Annals of the Entomol. Soc. of America* 89: 137-141.

Deka, M. & L. K Hazarika. 1998. Lifetime mating pattern of rice hispa, *Dicladispa armigera* Oliver (Coleoptera : Chrysomelidae). *Crop Research* 16 (1): 253-256.

Rao, P. S. P., P. Israel & Y. S. Rao. 1971. Epidomiology and control of the rice hispa, *Dicladispa armigera* Oliver. *Oryzae* 8 (2): 345-389.

Thakur, A. K., N. P. Kashyap, S. F. Hameed & S. M. Suri. 1979. Unusual occurrence of rice hispa on rice in Himachal Pradesh. *IRRN*. 4 (6): 7

Tiwari, S. K. & O. P. Dubey. 2001. Mating behaviour of rice hispa, *Dicladispa armigera* Oliver (Coleoptera : Chrysomelidae). *Crop Research* 21 (1): 109-111.

37

Response of Pure Silkworm, *Bombyx mori* L. Breeds to Feeding Methods and Feeding Sequences of Mulberry Varieties

CHINNASWAMY, K.P., CHANGALARAYAPPA
S. FATIMA, K. RASHMI AND N. JAGADISH

ABSTRACT

Feeding response of Pure Mysore, NB_4D_2 and NB_{18} breeds to chopped leaf and shoot feeding methods with Mysore local, M_5, and S_{54} mulberry varieties from *chawki* stage to late age with one variety and in feeding combination sequence of one to *chawki* and another to late age stage in direct combination and in reverse combination was assessed with respect to fifth instar larval weight, effective rate of rearing (ERR), cocoon weight and fecundity. Shoot feeding method recorded higher larval weight (305.25 g), 83.66 percent ERR, cocoon weight of 1.54g and maximum number of eggs (503.12) compared to chopped feeding method. NB_{18} responded with high larval weight (382.05g), and on par with NB_4D_2. Pure Mysore recorded significantly higher (87.65%) ERR to NB_{18} (77.58%). Significantly maximum single cocoon weight (1.83 g) was found in NB_{18}, than NB_4D_2 (1.82g). The breed NB_{18} laid maximum number of eggs (541.35), which was significantly higher than that of NB_4D_2 (530.96). Feeding sequence of M_5 + Mysore local resulted in maximum larval weight (308.91g). The feeding of S_{54} + Mysore

Local recorded higher ERR (86.00%) and was on par with Mysore local + S_{54} (84.4 %.). S_{54} + Mysore local sequence recorded higher cocoon weight (1.56g) over other sequences. The feeding sequence M_5 + Mysore local resulted in significantly more number of eggs (522.75) than with other feeding sequences. NB_4D_2, NB_{18} and PM silkworms responded very well to shoot feeding method with M_5 + Mysore local sequence in respect of egg yield and other traits.

INTRODUCTION

Mulberry is the sole food plant of the silkworm, *Bombyx mori* L. It derives nutrients required for its growth from mulberry leaves. Quality of the mulberry leaves has direct bearing on the growth and development of the silkworms and cocoons produced. Mulberry leaves with more moisture and protein content and less fiber are better for silkworm growth. Chemical compositions of leaves vary with varieties quantitatively and qualitatively. The importance of quality mulberry leaves on growth, development, health and silk production was stressed by Yokoyama (1963). Mulberry varieties have been reported to influence growth and development of silkworm (Krishnaswami *et al.*, 1970; Tayade *et al.*, 1988)

Commercial egg production depends on the quality of seed cocoons produced by the seed rearers. Sericulturists in recent years have been growing more than one mulberry variety for leaf production and rearing silkworm. It is well known that nutritive value of each variety differs. It is reported that, some varieties are best suited for early age and some are for late age worm rearing, depending upon the silkworm breed. However, adequate information is lacking on when, which variety and in which sequence to feed silkworm for efficient utilization of the available resources. In order to understand the feeding response of the different silkworm breeds to different varieties and method of feeding, a study was undertaken to evolve the best feeding sequence, method and breed performance with respect to grainage parameters.

MATERIAL & METHODS

Pure silkworm breeds of multivoltine and bivoltines were used for the study. Eggs of Pure Mysore breed were obtained from multivoltine seed station, Government silk farm, Bilidevalaya, Kunigal Taluk, Bangalore district. Eggs of bivoltine breeds namely NB_4D_2 and NB_{18} were procured from National Silkworm Seed Project, Bangalore.

Mulberry leaves of Mysore local, M_5 (Kanva-2) and S_{54} varieties raised under irrigated condition were used as feed for silkworms.

Adopting shoot and chopped method of feeding, silkworms were reared. Each silkworm breed was fed by three mulberry varieties separately, throughout the larval period. Further, two varieties, one in early stage and other in the later stage was fed in a feeding sequence. A total of six feeding sequences were performed for each breed. Thus, for three breeds, 18 feeding sequences besides, nine sole feedings amounted to 27 treatments.

For each treatment three replications were maintained with one DFL each. The treatment details are as follows.

T_1 : NB4D2 reared on local mulberry variety throughout.

T_2 : NB4D2 reared on local mulberry variety up to third instar and fourth and fifth instar on M_5 variety.

T_3 : NB4D2 reared on local mulberry variety up to third instar and fourth and fifth insatr on S_{54} variety.

T_4 : NB4D2 reared on M_5 mulberry variety throughout.

T_5 : NB4D2 reared on M_5 mulberry variety up to third instar and fourth and fifth instar on local variety.

T_6 : NB4D2 reared on M_5 mulberry variety up to third instar and fourth and fifth insatr on S_{54} variety.

T_7 : NB4D2 reared on S_{54} mulberry variety throughout.

T_8 : NB4D2 reared on S_{54} mulberry variety up to third instar and fourth and fifth insatr on local variety.

T_9 : NB4D2 reared on S_{54} mulberry variety up to third instar and fourth and fifth insatr on M_5 variety.

T_{10} : NB18 reared on local mulberry variety throughout.

T_{11} : NB18 reared on local mulberry variety up to third instar and fourth and fifth instar on M_5 variety.

T_{12} : NB18 reared on local mulberry variety up to third instar and fourth and fifth insatr on S_{54} variety.

T_{13} : NB18 reared on M_5 mulberry variety throughout.

T_{14} : NB18 reared on M_5 mulberry variety up to third instar and fourth and fifth instar on local mulberry variety.

T_{15} : NB18 reared on M_5 mulberry variety up to third instar and fourth and fifth insatr on S_{54} variety.

T_{16} : NB18 reared on S_{54} mulberry variety throughout.

T_{17} : NB18 reared on S_{54} mulberry variety up to third instar and fourth and fifth insatr on local variety.

T_{18} : NB18 reared on S_{54} mulberry variety up to third instar and fourth and fifth insatr on M_5 variety.

T_{19} : PM reared on local mulberry variety throughout.

T_{20} : PM reared on local mulberry variety up to third instar and fourth and fifth instar on M_5 variety.

T_{21} : PM reared on local mulberry variety up to third instar and fourth and fifth instar on S_{54} variety.

T_{22} : PM reared on M_5 mulberry variety throughout.

T_{23} : PM reared on M_5 mulberry variety up to third instar and fourth and fifth instar on local mulberry variety.

T_{24} : PM reared on M_5 mulberry variety up to third instar and fourth and fifth instar on S_{54} variety.

T_{25} : PM reared on S_{54} mulberry variety throughout.

T_{26} : PM reared on S_{54} mulberry variety up to third instar and fourth and fifth instar on local variety.

T_{27} : PM reared on S_{54} mulberry variety up to third instar and fourth and fifth instar on M_5 variety.

Treatment sequences

1. Mysore local +M_5 + S_{54}
2. S_{54} + M_5
3. M_5+ S_{54}
4. Mysore local + M_5
5. M_5 + Mysore local
6. Mysore local + S_{54}
7. S_{54} + Mysore local

The Silkworm rearing room was cleaned and washed with 1% bleaching powder solution. The rearing equipments were also cleaned and disinfected with bleaching powder solution and then dried under sun. Later, the equipments were kept inside the rearing room and disinfected with 2% formalien @ 800ml/ 10cm^2. The airtight condition was maintained for 48 hrs and then the room was opened for aeration (Narasimhanna, 1985).

Eggs of Pure Mysore, NB_4D_2, NB_{18} were procured from authorized sources and were incubated in disinfected room. The egg cards were arranged in wooden trays over a paraffin paper in a single layer and wet sponge stripes were placed all along the edge of the rearing tray and covered with paraffin paper to ensure the required microclimate for proper embryo development. At the egg pigmentation stage, eggs were subjected to dark treatment to ensure uniform hatching of eggs.

Brushing of silkworm intended to be reared with chopped method of feeding was done when 90 per cent of eggs were hatched according to Krishnaswami (1978). Silkworms required to be reared with shoot feeding were brushed in the following way:

When 90 per cent of the eggs hatched, the egg cards were placed on the shootlets of the mulberry and ensured that the worms crawled onto shootlets. However, the left over, if any, were made to fall onto the bed by gentle tapping or with the help of feather. Daily 4 feedings were given in both the methods of feedings and reared. Required number of worms were drawn and replaced as and when required to record observations.

Ripened worms of each treatment were mounted separately on bamboo mountages for spinning. The cocoons spun were harvested on 6th day of spinning. Observations recorded are: larval weight, effective rate of rearing, cocoon weight and fecundity.

RESULTS & DISCUSSION

Fifth instar larval weight (g/100 larvae) exhibited non-significant variation due to feeding methods. Shoot feeding method recorded significantly higher larval weight (305.246 g) compared to chopped feeding method (305.237 g). Fifth instar larval weight varied significantly among the silkworm breeds. NB_{18} breed recorded maximum larval weight (382.046 g), which was on par with NB_4D_2 (381.561 g) and significantly higher than PM (152.118 g). M_5 + Mysore local feeding sequence recorded maximum larval weight (308.909 g) which was on par with Mysore local + S_{54} (307.975 g) and S_{54} + Mysore local (308.227 g) and significantly higher than remaining feeding sequences viz., M_5 + S_{54} (303.698 g), Mysore local (229.74 g) and Mysore local + M_5 (297.939 g). Interaction effect of feeding methods, feeding sequences and silkworm breeds did not show any significant influence on larval weight. However, NB_4D_2 reared on M_5 + Mysore local with shoot feeding sequences recorded maximum larval weight (383.911 g), while PM reared on Mysore local mulberry variety with chopped feeding method recorded minimum larval weight (139.109 g) (Table 1).

Table 1. Effect of feeding methods, feeding sequences on fifth instar larval weight of pure silkworm breeds (g/100 larvae)

Feeding sequnces	NB_4D_2		NB_{18}		PM		Mean
	Chopped	Shoot	Chopped	Shoot	Chopped	Shoot	
1. Mysore local	379.80	380.14	379.61	379.91	139.10	139.91	299.74
2. M_5	380.66	380.62	383.59	383.89	146.73	146.66	303.69
3. S_{54}	381.23	381.22	381.75	381.54	150.83	151.61	304.70
4. S_{54} + M_5	381.04	380.34	380.06	379.88	148.86	148.67	303.14
5. M_5+ S_{54}	380.51	380.31	379.44	379.13	153.85	153.71	304.49
6. Mysore local + M_5	382.21	382.53	383.63	383.87	152.75	152.61	297.93
7. M_5 + Mysore local	382.28	382.54	383.10	383.40	159.12	158.91	308.22
8. Mysore local + S_{54}	383.46	381.54	383.42	383.83	157.86	157.71	307.97
9. S_{54} + Mysore local	383.66	383.91	383.20	383.50	159.49	159.66	308.90
Mean	381.65	381.46	381.98	382.11	152.07	152.16	
Mean	381.56		382.04		152.11		
Method Mean							
Chopped feeding				305.23			
Shoot feeding				305.24			

Factors/interactions effects	S.E+_	CD at5 %
Feeding methods (A)	22.65	64.19
Silkworm breeds (B)	27.74	78.62
Feeding sequences (C)	48.05	136.18
Interactions(AXB)	39.23	NS
(AXC)	67.96	NS
(BXC)	83.23	NS
(AXBXC)	117.71	NS

Effective rate of rearing varied significantly among the silkworm breeds. Pure Mysore breed recorded significantly higher effective rate of rearing (87.65 %) compared to NB_4D_2 (78.01%) and NB_{18} (77.58%). Among bivoltines NB_{18} and $NB_4D_{2,}$ effective rate of rearing was statistically non significant. Effective rate of rearing varied significantly due to feeding methods. Shoot method of feeding recorded an ERR of 83.66 per cent which was significantly higher than ERR of chopped method of feeding (78.49%). Feeding sequences had significant influence on the ERR. The feeding sequence of S_{54} to *chawki* worms + Mysore local to late age worms resulted in maximum ERR (86.00%) which was on par with feeding sequence of Mysore local + S_{54} (84.40%), M_5+ Mysore local (84.34%), Mysore local + M_5 (83.85%) and significantly higher than remaining feeding sequences viz., M_5 + S_{54} (78.97%), S_{54} +M_5 (80.16%), S_{54} (78.40%), M_5 (77.70%) and Mysore local (76.26%). ERR did not vary significantly due to the interaction effect of silkworm breeds, feeding methods and feeding sequences. However, maximum ERR (94.796%) was recorded with PM reared in sequence of S_{54} + Myosre local with shoot method of feeding and minimum ERR (69.18%) was recorded with NB_{18} reared on Mysore local in chopped method of feeding (Table 2).

Silk cocoon weight varied significantly among the silkworm breeds. Maximum single cocoon weight (1.831 g) was found in NB_{18}, which was significantly higher than that of NB_4D_2 (1.824 g) and PM (0.946 g). Single cocoon weight varied significantly due to the feeding methods. In the shoot feeding method, cocoon weight was 1.542 g, which was significantly higher than that of chopped method of feeding (1.525 g). Cocoon weight varied significantly due to the feeding sequences. S_{54} + Mysore local sequence recorded higher cocoon weight (1.556 g) which was significantly higher than that of other feeding sequences. Viz., M_5 + Mysore local (1.551 g), Mysore local + S_{54} (1.545 g), M_5 + S_{54} (1.533 g), Mysore local + M_5 (1.524 g) and S_{54} + M_5 (1.526 g), S_{54} (1.531), M_5 (1.522 g) and Mysore local (1.514 g). The interaction effect of silkworm breeds, feeding methods and feeding sequences had significant influence on the cocoon weight. Maximum cocoon weight (1.871 g) was recorded in NB_{18} breed reared on S_{54} + Mysore local in shoot feeding method and superior to all other interaction effects (Table 3).

Fecundity varied significantly among the silkworm breeds. The breed NB_{18} laid maximum number of eggs (541.346), which was significantly higher than that of NB_4D_2 (530.961) and PM (417.856). Further, the fecundity of NB_4D_2 and PM also differed significantly.

Table 2. Effect of feeding methods, feeding sequences on effective rate of raring of pure silkworm breeds

Feeding sequnces	NB_4D_2		NB_{18}		PM		Mean
	Chopped	Shoot	Chopped	Shoot	Chopped	Shoot	
1. Mysore local	70.19	75.99	69.18	76.22	75.99	86.26	76.26
2. M_5	70.82	78.45	71.85	77.89	78.45	88.04	77.70
3. S_{54}	72.62	79.14	69.44	78.43	79.14	88.81	78.40
4. S_{54} + M_5	74.40	78.76	74.53	80.37	78.76	89.66	80.16
5. M_5+ S_{54}	72.91	78.14	72.36	79.35	78.14	89.15	78.97
6. Mysore local + M_5	79.30	81.74	79.39	82.21	81.74	92.11	83.85
7. M_5 + Mysore local	79.78	83.24	79.17	80.15	83.24	93.74	84.34
8. Mysore local + S_{54}	79.36	82.76	79.11	82.11	82.76	92.97	84.04
9. S_{54} + Mysore local	82.00	84.53	80.69	83.99	84.53	94.79	86.00
Mean	75.71	80.30	70.05	80.03	84.70	90.61	
Mean	78.01		77.58		87.65		
Method Mean							
Chopped feeding				78.49			
Shoot feeding				83.66			

Factors/interactions effects	S.E+_	CD at5 %
Feeding methods (A)	0.34	0.98
Silkworm breeds (B)	0.42	1.20
Feeding sequences (C)	0.73	2.08
Interactions (AXB)	0.60	NS
(AXC)	1.27	NS
(BXC)	1.03	NS
(AXBXC)	1.80	NS

Table 3. Effect of feeding methods, feeding sequences on cocoon weight of pure silkworm breeds

Feeding sequnces	NB_4D_2		NB_{18}		PM		Mean
	Chopped	Shoot	Chopped	Shoot	Chopped	Shoot	
1. Mysore local	1.79	1.82	1.80	1.82	0.91	0.93	1.51
2. M_5	1.80	1.82	1.80	1.83	0.92	0.94	1.52
3. S_{54}	1.81	1.83	1.81	1.83	0.94	0.95	1.53
4. S_{54} + M_5	1.80	1.82	1.80	1.83	0.93	0.94	1.52
5. M_5+ S_{54}	1.81	1.83	1.82	1.84	0.94	0.95	1.53
6. Mysore local + M_5	1.80	1.82	1.80	1.82	0.93	0.94	1.52
7. M_5 + Mysore local	1.83	1.84	1.84	1.86	0.96	0.96	1.55
8. Mysore local + S_{54}	1.83	1.84	1.83	1.85	0.95	0.95	1.54
9. S_{54} + Mysore local	1.83	1.85	1.84	1.87	0.97	0.96	1.55
Mean	1.81	1.83	1.82	1.84	0.94	0.95	
Mean	1.82		1.83		0.94		
Method Mean							
Chopped feeding				1.52			
Shoot feeding				1.54			

Factors/interactions effects	S.E+_m	CD at5 %
Feeding methods (A)	0.001	0.002
Silkworm breeds (B)	0.001	0.002
Feeding sequences (C)	0.001	0.004
Interactions (AXB)	0.001	0.003
(AXC)	0.002	0.005
(BXC)	0.002	0.006
(AXBXC)	0.003	0.009

The feeding methods had significant influence on fecundity. In shoot feeding method, significantly higher number of eggs (503.125) were laid than chopped feeding method (490.316). Feeding sequences had significant influence on fecundity. The feeding sequence, M_5 to *chawki* worms + Myosre local to late age worms resulted in significantly more number of eggs (522.749) than the eggs laid with other feeding sequences Viz., M_5 + S_{54} + Mysore local (506.205), Mysore local + S_{54} (514.945), Mysore local + M_5 (490.282), M_5 + S_{54} (488.239), S_{54} + M_5 (445.164), S_{54} (506.128), M_5 (494.684) and Mysore local (502.090). Fecundity did not differ significantly due to interaction effects of silkworm breeds, feeding methods and feeding sequences. However, maximum fecundity (572.035) was recorded in NB_4D_2 breed reared in feeding sequence of M_5 + Mysore local in shoot feeding method. Minimum fecundity (344.835) was recorded in PM reared on S_{54} + M_5 feeding sequence in chopped method of feeding (Table 4).

Shoot feeding method recorded significantly higher larval weight (309.24 g) compared to chopped feeding method (305.23 g). These results are closely comparable with those of Chikkavenkateshappa (1987) who recorded larval weight of 298.40 g in shoot feeding method and 269.21 g in chopped feeding method, and Krishnaswami *et al.*, (1970) who also reported maximum weight of 1.96 g for the worms MBD_5 when reared on twigs with leaves of local mulberry cultivars. The lower weight in chopped feeding method may be attributed to poor quality of the leaf as it could not maintain the freshness of leaf by retaining turgidity due to chopping of leaf and rendered the leaf bits to dry faster, as such the worms consumed comparatively less resulting in delayed larval development.

The effective rate of rearing (ERR) was higher in shoot feeding method (83.66%) compared to chopped feeding method (78.49%). The lower rate of rearing in chopped method of feeding was mainly due to high mortality of worms due to grasserie and flacherie disease in fifth instar in all treatments. On the other hand the per cent survival was more in the case of shoot method of feeding, indicating the condition unfavorable to disease progression. These findings are in agreement with reports of Krishnaswami *et al.*, (1970) who found maximum of 50.21 per cent of effective rate of rearing in case of worms reared on twigs with leaves as compared to entire leaves. Further, feeding silkworms on chopped mulberry leaves (Nistari) up to second instar and shoot feeding till ripening stage resulted in higher effective rate of rearing (96.96%) compared to 95.51 per cent in silkworm fed whole leaves from 3rd instar onwards (Anon,1986). Chikkavenkateshappa (1987) reported an ERR of 51.01 per cent in shoot feeding method and 30.15 per cent in chopped feeding method.

Table 4. Effect of feeding methods, feeding sequences on fecundity of pure silkworm breeds (numbers/DFL's)

Feeding sequnces	NB_4D_2		NB_{18}		PM		Mean
	Chopped	Shoot	Chopped	Shoot	Chopped	Shoot	
1. Mysore local	492.39	508.63	565.27	572.95	429.73	443.55	502.09
2. M_5	513.23	524.81	545.35	559.05	403.57	422.07	494.68
3. S_{54}	526.35	538.85	553.33	564.31	417.84	436.07	506.12
4. S_{54} + M_5	479.29	495.94	488.82	498.73	434.83	336.36	445.16
5. M_5+ S_{54}	512.37	521.81	518.32	528.82	415.79	432.31	488.03
6. Mysore local + M_5	524.82	538.31	515.80	527.30	411.08	424.37	490.28
7. M_5 + Mysore local	564.85	572.03	558.84	569.82	426.30	444.64	522.74
8. Mysore local + S_{54}	522.72	559.55	561.82	565.29	415.78	434.49	514.94
9. S_{54} + Mysore local	561.87	569.42	520.30	530.05	417.81	437.77	506.20
Mean	525.32	536.59	536.43	546.26	406.19	426.51	
Mean	530.96		541.34		417.85		
Method Mean							
Chopped feeding				490.31			
Shoot feeding				503.12			

Factors/interactions effects	S.E+_	CD at5 %
Feeding methods (A)	0.721	2.044
Silkworm breeds (B)	0.883	2.503
Feeding sequences (C)	1.530	4.336
Interactions (AXB)	1.249	3.540
(AXC)	2.164	NS
(BXC)	2.650	7.510
(AXBXC)	3.748	NS

Cocoon weight was found to be influenced by feeding methods. Shoot feeding method resulted in significantly higher cocoon weight (1.54 g) compared to chopped feeding method (1.52 g). The present trend of higher cocoon weight in the shoot feeding methods is in confirmity with Krishnaswami *et al.*, (1970) who reported maximum cocoon weight (0.95 g) in case of MBD_5 reared on mulberry leaves with twigs. The advantage is that, the average cocoon weight was comparatively high in case of silkworm supplied with entire shoots compared to other methods. Application of chopped leaves to silkworm (Nistari) up to second instar alternated with shoot feeding resulted in maximum (0.92 g) cocoon weight, which was significantly high compared to those of silkworm fed with plucked leaves (0.85g) (Anon, 1986). Chikkavenkateshappa (1987) also reported higher cocoon weight (1.62 g) with shoot feeding method compared to chopped feeding method.

Fecundity varied significantly due to silkworm breed. Among the silkworm breeds, NB_{18} gave significantly maximum fecundity (541.34) compared to NB_4D_2 (530.36) and PM gave fecundity of 417.85. These results are in confirmity with Dutta (1984) who observed 638 and 572 eggs per laying in NB_{18} and NB_4D_2 breeds, respectively. Narasimhanna (1985) reported higher fecundity (530) in NB_{18} compared to NB_4D_2 (512). Further, Narasimhanna (1985) reconfirmed that NB_{18} and NB_4D_2 did not differ significantly with respect to fecundity.

SUMMARY

The response of silkworm breeds to different varieties of mulberry and feeding method was very positive to most of the parameters and differed significantly. The silkworm breed NB_{18} recorded more larval weight (382.04g), cocoon weight (1.83g) and fecundity of 541.34. Both NB_{18} and NB_4D_2 breeds were on par with most of the parameters. Naturally, multivoltine breed response cannot be compared to that of bivoltine breeds.

Feeding methods significantly influenced the rearing parameters of all the breeds. Silkworms reared with shoot feeding method recorded more larval weight (305.24g), effective rate of rearing (83.66%), cocoon weight (1.54g) and fecundity (503.12) compared to chopped leaf feeding method.

Feeding sequences had significant influence on grainage parameters. Feeding sequence of S_{54} to *chawki* worms + Mysore local to late age worms resulted in higher larval weight (368.90g), Effective rate of rearing (84.34%), land cocoon weight (1.55g) whereas feeding combination M_5 to *chawki* worms + Mysore local to late age worms resulted in high fecundity (522.74) and the sequence found to be best for seed breed rearing for seed cocoon production and further, for egg production.

The interaction effect of silkworm breeds, feeding methods, and feeding sequences showed signifificant influence on various parameters. More cocoon weight (1.85g) recorded in NB_4D_2 reared on S_{54} + Mysore local with shoot feeding method. Thus interaction of NB_4D_2 reared on M_5+Mysore local with shoot feeding method was found to be best for egg, cocoon parameters and interactions of NB_4D_2 reared on S_{54}+ Mysore local with shoot feeding method was found best for cocoon parameters. Similarly, rearing of NB_{18} and PM on M_5 + Mysore local for egg parameters and S_{54}+Mysore local for cocoon parameters with shoot feeding method was found the best.

Shoot feeding of M_5 mulberry variety during *chawki* and Mysore local during the late age was found best for egg yield and feeding of S_{54} during *chawki* and Mysore local during late age rearing was found best for cocoon yield of NB_4D_2 and NB_{18} and PM silkworm breeds.

Referecnes

Anon 1986. Standardization of rearing technique of silkworm (*Bombyx mori* L.) during commercial seasons in the tropical plains of West Bengal with special reference to feeding behavior. Annual report, Central Sericulture Research & Training Institute, Berhampore. pp. 17-20.

Chikkavenkateshappa, K.M. 1987. *Evaluation of different methods of application of feed to mulberry silkworms in relation to the economics of cocoon production.* M.Sc. (Agri.) Thesis, University of Agricultural Sciences, Bangalore. 81pp.

Dutta, R.K. 1984. Improvement of silkworm races (*Bombyx mori* L.) in India. *Sericologia* 24: 293-402.

Krishnaswami, S. 1978. New technology of silkworm rearing. Central Sericulture Research & Training Institute, Mysore. 24pp.

Krishnaswami, S., M. Asan & T.P. Sridharan. 1970. Studies on the quality of mulberry leaves and silkworm cocoon crop production. *Indian J. Seric.* 9: 11-25.

Narasimhanna, M.N. 1985. Silkworm seed organization and seed technology for India during the seventh plan. *Indian silk.* 24(4-5): 32-35 and 39.

Tayade, D. S., M. D. Jawale. & P.K. Unchegaonkar. 1988. Evaluation of some improved strains of mulberry by feeding experiment under Marathwada condition. *Indian J. Seric.* 27: 92-94.

Yokoyama, T. 1962. *Synthesized Science of Sericulture.* Central Silk Board, Government of India, Bombay. 397pp.

38

Feeding Potential of *Amblyseius Herbicoloides* Mcmurtry and Moraes (Acari: Phytoseiidae) on Selected Tetranychid Pests

SHEEJA, U.M. AND N. RAMANI

ABSTRACT

Phytoseiid mites are potential predators of various phytophagous mites and insects. All the stages of the predator exhibited voracious feeding and active searching behaviour. The predatory habit of various instars of *Amblyseius herbicoloides* varied considerably with respect to the stage/ species of the prey mite. Generally, the eggs of the prey mites were proved as the most preferred food item for all stages of the predator, irrespective of the species difference. The most voracious stage was the adult female and the predatory efficacy decreased in the order of deutonymph > male> protonymph > larva. The maximum rate of consumption was observed in the case of eggs (74.2% on *Tetranychus ludeni* and 95.5% on *T. neocaledonicus*) and minimum was on adults of the prey mites (5% on *T. ludeni* and 9.2% on *T. neocaledonicus*).

INTRODUCTION

Phytoseiid mites possess an array of supreme adaptive features which often raise them to the level of potential predators of pest mites, and also insects to a certain extent. In an agricultural country like India, biological control using introduced antagonists is becoming an essential component for the safe and effective management of agro-ecosystems. As a result of cultural practices to increase crop yield coupled with the indiscriminate use of chemicals for pest-control, the country is experiencing increasing instances of pest outbreaks. In this context, spider mites deserve special mention as several species are reported to show secondary pest out break, showing considerable resistance to many pesticides. Even though the predator mites play an important role in suppressing spider mite population in varied ecosystems, their importance is not properly recognized in India, particularly in Kerala. In this context the present study aims to unveil the feeding potential and behavioural aspects of *Amblyseius herbicoloides* (Mcmurtry & Moraes), a phytoseiid predator on two notorious species of pest mites viz. *Tetranychus neocaledonicus* and *Tludeni*, infesting a wide variety of crops.

MATERIAL & METHODS

Collection of both the predator as well as the prey mites was carried out from vegetable gardens, agricultural fields etc. located in different sites distributed over 3 districts of Kerala viz., Kozhikode, Malappuram and Trichur. Successful rearing and maintenance of sufficient stock cultures of the prey mites in the laboratory were carried out following leaf flotation technique. For this, leaf discs (2cm x 2cm) were cut out from respective host plants and kept in petri dishes lined with water saturated cotton pads. Live stages of the mites were transferred carefully to the leaf discs with the help of moistened camel hair brush. The leaf discs were renewed every 2-3 days. Several culture sets were maintained in the laboratory as described above at a temperature of $30 \pm 1^{0}C$ and a humidity of $70 \pm 2\%$ to study the feeding preference of individual stages of *Amblyseius herbicoloides* towards different stages of *Tetranychus neocaledonicus* and *Tetranychus ludeni.*

The experiment was initiated with setting of five culture sets containing leaf discs with different stages of the prey. Each leaf disc harboured 50 individuals of each of the stages of the respective prey mites (adults 25 only). Five such experimental sets of prey for each stage of predator were prepared viz. (1) Predator female Vs Prey egg (2) Predator female Vs Prey larva (3) Predator female Vs Prey Protonymph (4) Predator female Vs Prey Deutonymph (5) Predator female Vs Prey adult. Similarly different stages of predator viz., male, female, larva, protonymph and deutonymph were released to each culture set containing different stages of prey mites. Observations on feeding potential as well as feeding preference of immature and adult stages of the predator were made thrice in a day. Data obtained on the above parameters were tabulated and presented through appropriate tables/graphs. Data obtained on feeding potential were subjected to statistical analysis using repeated measures ANOVA followed by Scheffer test.

RESULTS & DISCUSSION

1. *Feeding Activity of A. herbicoloides on Spider mites* : The feeding process of the predator was initiated by touching the prey with its tarsus I, then grasping it with its second pair of legs. The first pair of legs held over the prey and then retrieved from the prey and were found extended anteriorly. Meanwhile, the palps were directed ventrad and slightly laterad, touching the prey with the dorsal setae on palp tarsus and tibial setae. The predator then bent its gnathosoma ventrad until its longitudinal axis was perpendicular to the prey's cuticle. In this position, the angle between the predator's body axis and gnathosoma was approximately 90^0. The chelicerae were alternately protracted beyond the apex of the corniculi, for a distance at least equal to the length of the movable digit. While the fixed digit laid outside the cuticle, the movable digit cut the cuticle. As soon as the body cuticle of the prey got penetrated, expansion and contraction of the pharyngeal muscles of the predator were observed, accompanied by the movement of fluids from the body of the prey into the predator's gut. The body colour of the predator was found changing and the coloured body fluid of the prey could be clearly seen through the transparent body of the predator.

   The feeding process of the various life stages of *A. herbicoloides* was more or less similar and all the instars were found very actively wandering on the leaf bits in search of the prey. However, the time taken by the different stages of the predator to devour the life stages of the prey mite showed variation. Accordingly, 5-10 minutes could be recorded for consumption of adult/nymphs of prey species and 3-5 minutes for eggs. The mode of consumption of egg by the predator was more or less similar to that of feeding on nymphs and adults. However, the palps were found laid over the egg and chelicerae were found extended to the eggs. Several punctures were produced on the egg surface with the help of the extended chelicerae, through which the contents were withdrawn leaving behind the collapsed egg shell.

2. *Feeding Potential of Predator Life Stages on Spider mites*: Results of feeding studies revealed that *A. herbicoloides* preferred the immature stages of the prey irrespective of the difference in prey species. However, the rate of consumption of the different stages of the predator showed variation with respect to change in prey. Most of the life stages of *A. herbicoloides* showed a preference to the eggs of prey mites, thereby supporting the earlier findings on *A. bibens* (Blommers & Etten, 1975) and *A. fallacis* (Santos, 1975). Contrary to these, *phytoseiulus persimilis* showed more affinity to the adults and nymphs of *Tetranychus urticae* (Shehatha, 1973). *A. alstoniae* preferred larvae and protonymphs rather than other stages of *Eutetranychus orientalis* (Dhooria, 1981). The order of preference of the various stages of *A. herbicoloides* to life stages of prey mites followed a decreasing sequence as: Egg> Larva> Protonymph> Deutonymph> Adult. Maximum rate of consumption in *A. herbicoloides* was observed on the prey eggs and the lowest rate could be noted on prey adults. Thus, the feeding potential of *A. herbicoloides* closely followed that of *Typhlodromus occidentalis* which fed on *T. macdnanielli* (Croft & McMurtry, 1972). For the purpose of biological control, the preference of *A. herbicoloides* on prey egg could be considered as an advantage, because it could help to reduce the prey population well below the threshold level.

As the development progressed, the consumption rate increased in *A. herbicoloides*. Similar results were recorded in other phytoseiid predators like *Neoseiulus anonymus* (Mesa & Bellotti, 1986) and *A. cydnodactylon* (El-Benhawy *et al.*, 2000) when fed on *T. urticae*. The larva of *A. herbicoloides* was found feeding on the eggs and nymphs (protonymphs) of the prey species. It initiated feeding shortly after hatching as reported earlier in the case of *I denerans* (Takafugi & Chant, 1976.), *M. occidentalis* (Friese & Gilstrap, 1985) and in *A. californicus* (Tetsuo Gotoh *et al.*, 2004). However, in species like *Phytoseiulus persimilis* (Takafugi & Chant, 1976), *A. fustis* (Ezulike & Odebiyi, 1984) and *A. tetranychivorus* (Neelu Nangia & Channa

Basavanna, 1989), the larvae were reported to be at a non feeding stage. Such non-feeding larval trait was designated as an adaptation to avoid cannibalism in species which lay eggs close together (Chittenden & Saito, 2001). However, the feeding behaviour of larval stage in *A. herbicoloides* may serve to enhance the predatory potential of the species on pest mite species tested. No cannibalistic trait could be observed in the larvae of this species, but instead they showed varying rates of consumption of the prey mites. The consumption rates of larvae varied with respect to species variation and life stages of the prey. Several other species of phytoseiid predators like *A. barkeri* (Momen, 1995) also showed a similar trend on *T. urticae* and *Eriophyes dioscoridis*.

Table 1. Mean Daily consumption rate of *A. herbicoloides* on *T. ludeni* at 30±1°C and 70± 2% RH

Sex and growth stages of *A. herbicoloides*		Predation on various life stages of prey, *T. ludeni* (X + S.E)				
		Egg	Larva	Protonymph	Deutonymph	Adult
Adult	Female	37.1±0.56	32.5±0.55	28.4 ± 0. 71	4.9 ± 0.26	2.5 ± 0.32
	Male	25.1±0.67	18.8±0.56	16.3 ± 0.45	2.7 ± 0.28	1.4 ± 0.29
Nymphal Stages	Deutonymph	28.8±0.68	23.8±0.77	17.2 ± 1.10	3.3 ± 0.35	2.3 ± 0.20
	Protonymph	16.8±1.81	13.8±0.58	7.3 ± 0.57	0	0
	Larva	2.7 ± 0.35	1.9 ± 0.22	0	0	0

Table 2: Mean Daily Consumption Rate of *A. herbicoloides* on *T. neocaledonicus* at 30±1°C and 70± 2% RH

Sex and growth stages of *A. herbicoloides*		Predation on various life stages of prey (X + S.E) *T. neocaledonicus*				
		Egg	Larva	Protonymph	Deutonymph	Adult
Adult	Female	47.7 ± 0.55	38.9 ± 1.36	29.7 ± 0.55	7.4 ± 0.64	4.6 ± 0.62
	Male	41.9 ± 0.54	21.8 ± 1.16	18.3 ± 0.80	4.1 ± 0.41	2.6 ± 0.29
Nymphal Stages	Deutonymph	44.7 ± 1.25	27.6 ± 0.71	22.8 ± 0.69	5.7 ± 0.57	3.1 ± 0.36
	Protonymph	31.1 ± 0.75	17.9 ± 0.51	12.8 ± 0.61	0.7 ± 0.14	0
	Larva	5.3 ± 0.75	1.3 ± 0.25	0	0	0

Adults, especially adult females of *A. herbicoloides* consumed more number of prey mites than the other stages. The highest rate of predation by the female was observed during the oviposition period. The protonymph and deutonymph of this predator had higher feeding potential than the larvae, but the most voracious stage recognized during the study was the adult. The maximum daily consumption rates of adult female on *T. ludeni* were 37.1 eggs, 32.5 larvae, 28.4 protonymph, 4.9 deutonymph & 2.5 adults. An increased consumption rate of 47.7 eggs, 38.9 larvae, 29.7 protonymphs, 7.4 deutonymphs and 4.6 adults could be observed on a diet of *T. neocaledonicus*. The maximum rate of consumption was observed in the case of eggs, reaching 74.2% *in T. ludeni* and 95.5% on *T. neocaledonicus* and minimum was on adult prey (5% on *T. ludeni* and 9.2% on *T. neocaledonicus*). Similar results were reported in *Neoseiulus anonymus* on *T. urticae* and *E. dioscoridis* (Mesa & Bellotti, 1986). These differential feeding on *T. ludeni* and *T. neocaledonicus* showed that *A. herbicoloides* prefer *T. neocaledonicus* than *T. ludeni*. The feeding efficacy of the deutonymphal stage of *A. herbicoloides* was comparatively higher than that of the adult male. But the data obtained on the consumption rate of male when analysed statistically showed a significant value, thereby confirming its predatory potential. However, the predatory potential of *A. herbicoloides* decreased in the sequence: Female> Deutonymph> male> Protonymph> Larva. The rate of consumption on both the prey mites by the predator/day was found to be statistically significant at 0.01 level. This clearly substantiates the potential of *A. herbicoloides* in suppressing the populations of both species of spider mites tested.

CONCLUSION

The results of feeding studies clearly established *T. neocaledonicus* as the most preferred prey for *A. herbicoloides*, as observed during the current investigation. Thus the species could be confirmed as one of the most efficient predator of both *T. neocaledonicus* and *T. ludeni*, two major pests of vegetable crops, based on the significant results drawn through statistical analysis comprising repeated measures ANOVA and Scheffer Test.

References

Croft, B. A. & M. C. Murtry. 1972. Minimum release of *Typhlodromus occidentalis* to control *Tetranychus mcdanieli* on apple. *Journal of Economic Entomology* 65: 188-191.

Shehata, K.K. 1973. Prey type and density and their effect on the potentiality of *Phytoseiulus persimilis* Athias-Henriot (Acarina: Phytoseiidae). *Biologia Czechoslovakia* 28.8: 619-626.

Santos, M.A. 1975. Functional and numerical responses of the predatory mite, *Amblyseius fallacies* to prey density. *Environmental Entomology*.6: 989-992.

Blommers, L. & J. V. Etten.1975. *Amblyseius bibens* (Acarina: Phytoseiidae), a predator of spider mites (Tetranychidae) in Madagascar. *Entomologia Experimentalis et Applicata* 18.3: 329-336

Takafuji, A. & D. A. Chant. 1976. Comparative studies of two species of predaceous phytoseiid mites (Acarina: Phytoseiidae), with special reference to their responses to the density of their prey. *Researches on Population Ecology* 17.2: 255-310.

Dhooria, S. 1981. Feeding behaviour of predatory mites, thrips and beetles of the citrus mite, *Eutetranychus orientalis*. *Acarology Newsletter* 10: 4-6.

Ezulike, T. O. & J. A.Odebiyi. 1985. Life history of *Amblyseius fustis* (Pritchard and Baker), an indigenous predator of the cassava red mite, *Oligonychus gossypii* (Zacher) in South Western Nigeria. *Insect Science and its Application* 6.2: 193-197.

Friese, D. D. & F. E. Gilstrap. 1985. Prey requirements and developmental times of three phytoseiid species predaceous on spider mites. *South Western Entomologist* 10.2: 83-88.

Mesa, C. N. C. & C. Bellotti. 1986. Life cycle and feeding habits of *Neoseiulus anonymus*, predator of tetranychid mites in cassava. *Revista Colombiana de Entomologia*. 12.1: 54-66.

Nangia, N. & G.P. Channa Basavanna. 1989. Feeding potential of *Amblyseius tetranychivorus* (Acari: Phytoseiidae), an indigenous predator on selected tetranychid and tenuipalpid mites. *Indian Journal of Acarology* 10.1 & 2: 75-81.

Momen, F.M. 1995. Feeding, development and reproduction of *Amblyseius barkeri* (Acarina: Phytoseiidae) on various kinds of food substances. *Acarologia* 36.2: 101-105

El-banhawy, E. M., S. A. A. Amer & S. A. Saber. 2000. Development and reproduction of the predaceous mite, *Amblyseius cydnodactylon* on different prey species: Effect of plant leaf texture on the behaviour and reproduction of the predator. *Zeitschrift fur Pflanzenkrankheiten und Pflandzenschtuz* 107.2: 218-224.

Chittenden, A.R. & Y. Saito. 2001. Why are there feeding and non feeding larvae in phytoseiid mites (Acari: Phytoseiidae). *Journal of Ethology* 19.1: 55-62.

Tetsuo, G., Y. Koichi. & Katsuhiko Mori. 2004. Effect of temperature on the life history of the predatory mite *Amblyseius* (*Neoseiulus*) *californicus* (Acari: Phytoseiidae). *Experimental and Applied Acarology* 32:15-30.

39

Foraging and Feeding Behavior of Indian Crested Porcupine (*Hystrix indica* Kerr) in Wild and Cultivated Areas of Western Ghats of Karnataka

CHAKRAVARTHY, A.K., A.C. GIRISH
AND N.R. PRASANNA KUMAR

ABSTRACT

As a result of declining forest cover and using of forest cover for cultivation in Karnataka wild animals like porcupine are invading and utilizing cultivated ecosystems for foraging and feeding. Distribution of porcupine burrows is based on the distribution of natural food resources. Cultivation will change this distribution of food resources which in turn will affect porcupine population in a particular area. Individuals in a burrow form a family and the individuals in adjacent burrows in a given patch will form a group. These individuals of a group in a patch form a community/clan/colony. Porcupines show territoriality. The behavior of porcupine in wild differs from the behavior exhibited in the cultivated ecosystem. But, both behaviors depend on food availability. Account of such differential behaviour in cultivated and wild patches is provided in this paper.

INTRODUCTION

Porcupines are herbivorous animals feeding on certain roots, bulbs and succulent tubers and prefer the bark of certain tree species. They also feed on ripe fallen fruits (Roberts, 1997). Nowak (1991) reports that porcupines may cover 15 km in one night in search of food but Kingdon (1974) writes that they cover 30 km before returning to their burrow the next morning. In Baluchistan, porcupines excavate bulbs of *Eremurus aurantiacus* and in the South west Punjab they prefer bark of trees, *Melia azedarach, Moris alba and Mangifera indica.* They also debark trees with thick and rough bark such as *Dalbergia sisoo.* Porcupine foraging is very destructive to agricultural crops like maize, sugarcane, potato, sweet potato and ripe melons (Roberts, 1977).

Food preferences of porcupines has received considerable attention because, their winter feeding may permanently damage or kill trees (Gabrielson, 1928; Brander, 1973; Speer & Dilworth, 1978). Tenneson & Oring (1985) studied the damage inflicted by porcupine (*E. dorsatum*) and their habitat use at two locations in Itasca state park in North Western Minnesota. Porcupines preferred trees for debarking in proportion to their availability. Large trees were more prone to damage (15 cm dbh) and the amount of damage increased with greater tree size. Overall, tree species used by porcupines differed between two locations. The proportion of porcupine damaged trees did not increase with the decreasing species abundance which was contrary to Westoby's (1974) findings. In the present study efforts were made to record feeding and foraging of porcupines in natural and cultivated areas of Western Ghats of Karnataka where the population of the porcupine is declining

MATERIAL & METHODS

To study the foraging behaviour of porcupine, Telonics Telemetry system was utilized. Trained locals and tribals of Subramanya, Dakshina Kannada district were employed to capture the animals. This was done after observing porcupine's used runways. A trench (3′ x 3′x 4′) was dug in its runway. The trench was covered at the top with coconut fronds and grass for trapping it in the pit. Animals trapped were secured with the help of a nylon rope, immobilized with 20 mg/individual of xylazine hydrochloride, mixed with 10 mg of Ketamine hydrochloride per kg of body weight, injected in the lower back by a jab stick (Alkon, 1984; Pigozzi, 1987). The animal was covered with ice cubes wrapped in gunny bags to reduce the increased temperature as a result of injecting the chemical. The animals were weighed, blood sampled, earmarked and sprayed with enamel organic dyes on the hindquarters. Animals weighing less than four kg were classified as juveniles. Two adult females were used to fit collars. They were fitted with soft, sturdy canvas radio collars (Telonics, USA). Each transmitter was on a different frequency, between 150-850 and 151-425 MHz. The total weight of each radiocollar was 65 g. A loop antenna, built inside the collar, and a three element potable Yagi aerial, together with a receiver (Telonics, USA) formed the telemetry system. Data recorded in the receiver was not considered for the first few days until the animal acclimatized with the radiocollar and overcame the effect of sedation, which may last for several days (Alkon, 1984).

Each porcupine was monitored once a week (24 hours and 18 hours, dusk to dawn, on alternative weeks), from January to December 2004. At each location data were recorded for continuous 15 minute intervals throughtout the night and every 30 minutes through out the day from two or when necessary, three reference stations. The porcupines were assumed inactive when inside the burrow, but sometimes they rested also on the surface, under vegetation cover. Duration of activity was measured as the difference between the times of onset and termination of surface activity. Nocturnal activity was measured from 60 min. after sunset to 60 min. before sunrise. Crepuscular activity included 60 min. before and after sunset and 60 min. before and after sunrise. Diurnal activity ranged from 60 min after sunrise to 60 min before sunset. We have called dark hours as the sum of nocturnal and crepuscular hours. Based on the strength and frequency of the beeps, the data recorded in the receiver was interpreted to arrive at the patterns of habitat use by the animals and by visual observations for selected behavioral traits. The observations recorded with head searchlights along with knowledgeable forest staff and tribals were corroborated and validated with the information recorded through the telemetry system.

Correlations were tested through Spearman's rank correlation coefficient (Siegel, 1956). The significance of differences between two porcupines (%) was assessed according to Hald (1967).

RESULTS & DISCUSSION

The details of the two animals trapped for radio tagging are furnished in Table 1. The weight of the female porcupines ranged from 14 to 16 kg (n=4). Head and body length of porcupine varied from 70-83 cm and tail length from 10-14 cm. The nocturnal activity of the tagged animals ranged from 10h to 11h. It varied depending on the season. The animals were found to leave the burrow when sun sets and return early in the morning.

Table 1. Descriptive parameters of radio tagged porcupines

Parameters	F1	F2
Weight of the animal (kg)	14.5	15.3
Body and Head length (cm)	79	83
Tail length (cm)	10	12
Nocturnal activity (h)	134 45	134 20
Diurnal activity (h)	10h 15′ (N=15) 15′	10h 40′ (n+15) 25′

F_1 and F_2 refer to test animals, both were female.

Night surface activity of the radio tagged porcupine depended on the season. As the dark hours decreased (February to May) the activity of porcupine also declined. Porcupine was found more active per hour of the dark night during winter and rainy seasons compared to summer. The availability of food due to fresh vegetative growth during rainy season made porcupine active during night. During summer due to hot conditions they took rest intermittently in the middle of their foraging bout. Diurnal activity of porcupine was also observed and it was occasional and was not frequent. Daylight movement were usually confined to surroundings of the burrows (n=30; 68.1% within 100m from the den). Usually porcupines rest inside the burrows during day time but sometimes if too hot, they will rest in the bushes near the burrows.

Observations were recorded at the time of leaving the burrow and return to the burrow daily (N=8) at Subramanya, in coastal Karnataka. The habitat covered and the activities of the porcupine are presented in Table 2. During rains the animal took shelter outside and returned to the burrow next day. Similar activity was observed in B.R. forest area also where the animals took shelter under thick bamboo bush when it rained. Time of exit and entry to the burrows was in the dark hours only i.e. after sunset and before sunrise, respectively. Porcupines mostly left the burrows after sunset (73% of cases, n= 55), particularly in colder season (87%, n=35). This habit decreased significantly (P<0.001) in the warmer months (58.8*%, n= 35). Their activity tended to stop before sunrise (78.9% of the cases, n=60). Porcupines showed a remarkable continuity in their activity bouts. In only 16.3% of the monitored nights they interrupted their activity with a resting period, before finally returning to the burrows. These breaks could last several hours (mean duration; 2h 30" range: 15" to 4h 30"), occurring mainly during March–April (summer months) (75.5%, n=25 nocturnal inactive duration).

Table 2. Length of foraging bouts of porcupines at Subramanya

Date	Weather condition	Leaving (p.m)	Returning (a.m)	Foraged duration (hr)	Habitat covered by the animal	Activity
5-11-03	Cool	7.00	4.30	9.30	Ravines, sewage canal, undulating land	Scaning (S) Foraging (F)
6-11-03	Cool	6.20	4.45	10.25	-do-	S,F
7-11-03	Cool	6.35	4.00	9.25	-do-	S,F
8-11-03	Intermittnt rains	6.25	4.30*		-do-	S,F
9=11-03	Cool	6.30*	1.00		-do-	S,F
10-11-03	Cool	6.45	3.30	8.45	-do-	S,F
11-11-03	Cool	6.1o	12.30	6.20	-d0-	S,F
12-11-03	Cool	6.30	3.00	8.30	-do-	S,F

*The female that returned was not the one residing inside the burrow. It came alone and returned when its partner came near the burrow next day

Corsini *et al.*, (1995) by radio tagging porcupine, also recorded that the animals were active during night at an average of 9h 30″. The foraging path was recorded daily for 15 days by following the animal without disturbing it. The effect of rains on porcupine foraging activity at Mysore Zoo is presented in Table 3. The activity was observed at different timings during the night. In B.R. forest tracts and other areas, porcupine covered upto 2-3 km per night for foraging.

Porcupines grunt and rattle quills as they shuffle around at night. When they encounter other animals, they raise and fan their quills, more than doubling their apparent size. If still bothered, they stamp their feet, whirrl their quills and finally charge backwards towards the enemy.

Roberts (1977) has recorded porcupine feeding on several agricultural crops. Bruno & Riccardi (1995) have also recorded porcupine's wide food range by studying fecal pellets throughout the year in a rural hilly area in central Italy. These studies indicate porcupine as a generalist forager. Porcupines preferred to forage during night in ravines, paths under culverts and under the shade of trees like rain tree (*Samanea saman*), species of *Cassia* and *Acacia.* Bark feeding was seasonal. Porcupines feeding on the bark may be to sharpen their incisors, aid in digestion and also to get nutrition. Dodge (1982) also recorded seasonal use of bark. Similarly use of bark for nutrient supplementation was observed by Banfield (1974).

Table 3. Effect of rains on porcupine foraging activity

Week	Weather conditions	Activity	Average no. of animals sighted at different timings at night per day (male + female)				
Oct. 1st week	Cool, drizzle of rain intermittently	F*, S*	1+1	2+2	-	1+1	4+3
Oct. 2nd week	Rains during night	F, S	1+2	-	-	-	1+2
Oct.3rd week	Heavy rains during night	No activity	-	-	-	-	-
Oct.4th Week	Rains during 11.00 pm to 5.00 am	F, S	1+0	-	-	-	1+0
Nov. 1st week	Heavy rains during night	F, S	-	2+2	-	-	2+2

*F-foraging, S-scaning. Activity of porcupines was observed in different timings during the night (1) 7.00 to 9.30 pm (2) 10.30 pm to 12.00 am (3) 1.00 to 3.00 pm (4) 4.00 to 7.00 am. Its activity in the night was recorded.

ACKNOWLEDGEMENTS

The authors thank Department of Science and Technology, New Delhi for financial assistance, Regional Research Station of Central

Plantation Crops Research Inistitute at Kidu for permitting observations in their campus, local farmers and tribal people for assistance and to Dr.Hossetti for comments on an earlier draft of this manuscript.

References

Alkon, P.U. 1984. Chemical restraint of Indian crested Porcupines (*Hystrix indica*). *Mammalia*. 48: 150-152.

Banfield, A.W.F. 1974. *The Mammals of Canada*, University of Toronto Press, Toranto, Canada. XXV + 438 pp.

Brander, R.B. 1973. Life history notes on the porcupine in a hardwood-hemlock forest in upper Michigan. *Michigan Academician* 5: 425-33.

Bruno, E. & C. Riccardi. 1995. The diet of the crested Porcupine *Hystrix cristata* L. 1758. *Michigan Academician* 5: 226-236.

Corsini, M. T., S. Lovari & S. Onnino. 1995. *The Mammals of Indo Malayan region. A systemic review*. Oxford University press, Oxford, UK.VIII + 488 pp.

Gabrielson, I.N. 1928. Notes on the habits and behavior of the porcupines in Oregon. *J.Mammal*.9: 33-38.

Hald, A. 1967. *Statistical theory with engineering application*. Wiley, New York.

Kingdon, J. 1974. *East African Mammals: An evolutionary Atlas, Vol. 2*. Academic press, New York. 404 pp.

Nowak, R.M. 1991. *Walker's Mammals of the world. Vols. 1* and 2. The John Hopkins University press, Baltimore, USA. 1629 pp.

Piggozzi, G. 1987. Immobilization of crested porcupines with xyline hydrochloride and ketamine hydrochloride. *J. Wildl. Manage*. 51 (1): 120-123.

Roberts, T.J. 1977. *The Mammals of Pakistan*. Ernest Benn Ltd., London/Tonbridge. 361pp.

Siegel, S. 1956. *Nonparametric statistics for the behavioral sciences*. Mc Graw-Hill, New York.

Speer, R. J. & T. G. Dilworth. 1978. Porcupine's winter foods and utilization in central New Brunswick. *Can. Field Nat*. 92: 271-274.

Tennesson, C & L. W. Oring. 1985. Winter food preferences of porcupines. *J. Wildl. Manage*. 49: 28- 33.

Westoby, M. 1974. *An analysis of diet selection by large generalist herbivores. Am. Nat.* 108:290-304.

40

Feeding Behaviour of Selected Insect Predators on Sugarcane Woolly Aphid, *Ceratovacuna lanigera* Zehnter (Homoptera Aphididae) in South Karnataka

MUNISWAMY GOWDA K. N.
A. K. CHAKRAVARTHY, L.VIJAY KUMAR AND PRASANNA KUMAR, N.

ABSTRACT

Sugarcane woolly aphid (SWA), *Ceratovacuna lanigera* is a plant sap sucking insect with a highly restricted host range. It feeds and breeds in colonies under the surface of sugarcane leaves. The aphid is covered with white woolly matter. Of the 11 natural enemies recorded on the aphid, *Dipha aphidivora* Meyr. (Pyralidae; Lepidoptera) and *Micromus igoratus* (Hemerobiidae; Lepidoptera) are important and amenable for innundative releases. *D. aphidivora* is a nocturnal feeder, recording a feeding potential of 35-40 aphids/day in Mandya. The total life cycle of the predator varied from 30-50 days under field conditions. *M. igoratus* is a diurnal feeder with a feeding potential of 18±2.2 aphids/day. The total life cycle of the predator lasted 20±2.80 days in Mandya under field conditions. The predator larvae fed intermittently on aphids. The hemerobiid larvae preferred to feed on aphids with less mealy matter while, *D. aphidivora* larvae preferred to feed on all stages of the woolly aphid. *Metarhizium anisopliae* and *verticilium* (fungal pathogens) did not drastically reduce SWA. Replicated field trials indicated that natural enemies help in sustaining sugarcane yields provided other management tools are also incorporated.

INTRODUCTION

Sugarcane is one of the most important commercial crops in the tropics and serves as source of sugar world over. India is one of the leading countries in sugar cultivation covering nearly four million ha with annual production of 265 million tonnes (Lingappa, 2004). The sugar production in Karnataka fell from 330 lakh tonnes in 2001- 02 to 158 lakh tonnes in 2003-04. The crop is attacked by about 228 insect and non- insect pests in India (David & Nandagopal, 1986). Sugarcane woolly aphid (SWA), *Ceratovacuna lanigera* Zeh. (Homoptera: Aphididae) was a minor or negligible pest occurring in tropical Asian regions. Basu & Banerjee (1958) reported the aphid for the first time in India from Coochbehar district of West Bengal. After one and a half decades, Ghosh (1974) reported aphids from North Eastern states and Uttar Pradesh. However, its pestiferous form was noticed after 25 years later in Maharashtra and Karnataka where instantaneous outbreak of SWA was noticed in 2002. Currently woolly aphid is a serious pest of sugarcane in Karnataka. Globally 40 natural enemies have been recorded, which include predators (30), parasitoids (06) and pathogens (04). In north Karnataka, Vidya Mulimani *et al.*, (2004) recorded 11 natural enemies on SWA. Patil (2004) also recorded 11 natural enemies including spiders from Kolhapur and Sanghvi regions. Of these, three predators were commonly found viz, *Dipha aphidovora* Meyr, *Micromus* spp. and syrphids. This paper deals with the behavioral aspects of the first two predators and two fungal pathogens on SWA in field and laboratory.

MATERIAL & METHODS

Field observations on the activity of the two predators, *Dipha* and *Micromus* were recorded in cane fields during 2003–04 in Mandya district (11^0 30^1 N to 13° $05^1 05^1$ E to 77° 45^1 and 76° 05^1) in southern Karnataka. To determine the distribution pattern of the predators, the vertical height of the sugarcane tillers in the field were divided into bottom, middle and top portions arbitrarily. Counts of predators in the above three levels were made at different intervals. Data were subjected to appropriate statistical analyses.

To determine the feeding potential, SWA infested leaf bits were placed in petridish (6 inch dia) over a wet cotton spread and covered on top with the same size blotting paper. The number of aphids consumed by an adult and a larva of the predator at definite intervals of time was determined. Observations were recorded in laboratory on the life history of the predators. The number of generations completed by the predators on SWA was also determined.

The infestation of SWA was assessed based on the leaf area covered by the aphid colony and expressed in terms of number of infested leaves. Thus, if 50% of undersurface of two leaves were covered by aphids, it is equivalent to one aphid infested leaf. The number of predators on the undersurface of all leaves of a clump were counted. Observations were recorded once in 15 days on five, randomly selected clumps/plot. The average of two observations formed the number of infested leaves and predators for each month. The aphid infestation and their predators were monitored for one year from May 2002 at Tarikere, Chikmagalur on Co. 8371 variety and from June 2003 at Ghandal, Mandya on the same variety.

To find the effect of two fungal pathogens in suppressing SWA, an experiment was laid out in split plot design with 4 treatments, 3 sub plot treatments with 4 replications each. The required quantities of Biomet (*Metarhizium anisopliae*) and Biovert (Verticillium) were mixed in one litre of water each separately and it was sprayed onto plant canopy by sprayer. While spraying, each clump was fully covered (on an average each clump consisted of about 7.8 canes, n =25 clumps) with the spray solution and one clump was treated as one replication/ treatment. The increase or decrease in aphid infestation after treatment was scored as follows: if the aphid infestation decreased or increased by one leaf, one point was subtracted or added. Accordingly, if the aphid infestation decreased or increased by half leaf 0.5 was subtracted or added. Higher scores indicated greater infestation. The control plot was maintained free of woolly aphids and their natural enemies. If any natural occurrence of the aphids and their predators was observed, the insects were transferred to plots of only woolly aphids and natural enemies to plots of woolly aphids + natural enemies plots. The data were subjected to appropriate statistical analyses.

RESULTS & DISCUSSION

The predator, Brown lacewings, *Micromus* spp. (Neuroptera: Hemerobiidae) was one of the common species found devouring SWA in all sugarcane growing areas of Karnataka. Larvae are predacious and resembled 'crocodile'. Gravid females of hemerobiids were observed laying dirty white eggs at lower portions of cane just above soil surface on either sides of lower leaf or on stem sheath. Incubation on an average lasted for 3.5 ± 0.50 (n=40) days. Newly emerged larvae were flattened; spindle shaped and underwent three instars. Larval duration, on an average, lasted 6.0 ± 0.78 (n=25) days. Fully grown larva was dark to blackish brown. Larvae were characterized by a pair of sickle shaped mandibles. They were voracious feeders. On an average, a maximum of 16–18 larvae per leaf were noticed feeding on SWA (n=30). Vidya Mulimani *et al.*, (2003) and Shekharappa (2003) in Dharwad recorded 3 to 4 folds more number of larvae/leaf.

Cocoons of hemerobiids were observed at bottom portions (< 0.5 m from soil) of canes, under the stem sheath or lower side of bottom leaves. The pupae were covered in a double cover; outer cover was silky white and inner cover was brown. Pupae were oval and dirty white to light brown.

Adults were brown, medium sized. Brown lacewings were nocturnal and were phototatic. Adults were also observed feeding on woolly aphids. Total life cycle lasted 20 ± 2.80 days under Mandya conditions during September–October. Feeding potential of adults under laboratory conditions on an average was 18 ± 2.0 SWA/day.

More number of hemerobiids were found at the bottom portions of canes compared to the middle and top portions (Table 1). There were statistical differences in number of hemerobiids between the top and bottom portions of canes. Hemerobiid larvae were found feeding on aphid nymphs with no or less woolly matter and were highly selective. The larvae appeared to prefer aphids with entire body covered with less mealy matter. Ants were not found interfering in the predator activity of hemerobiids. In general, fourth instar aphid nymphs were preferred by the hemerobiids.

Early instar larvae of hemerobiid fed on first and second instar aphids and they were more active than grown-up ones. Grown up larvae of hemerobiids fed on grown up nymphs of aphids. The predatory larvae initiated feeding from posterior end and at the ventral surface. Continuous observations in laboratory revealed that the hemerobiid larvae fed on aphids at the average rate of 1 aphid/10 to

20 min. Field observations were confirmed under laboratory in Petri plates in a replicated trial. On an average, one hemerobiid larva fed on one aphid in 15 to 20 minutes. After devouring a larva, the predator remained motionless and without feeding for 10-15 min. Thus the hemirobid larva did not continuously feed on aphids (Fig 1). The predator preferred to feed under hidden dark places or niches on the plant. On an average, a larva fed 24 to 30 aphids/day (n=28) in Mandya.

Table 1 Vertical distribution of *Micromus* spp. larvae on cane leaves

Vertical category	Number of *Micromus* larvae/leaf*
Bottom (0-0.4m)	1.25 (7.57)
Middle (>0.04-0.8m)	0.85 95.25)
Top (> 0.8m)	0.28 (3.03)
Sem ±	0.785
CD at 5%	2.421

* Mean of 7 replications: 10 clumps/replicate: values in the parentheses are vx+ 0.5 transformed values.

The predator, *Dipha aphidivora* Meyrick (Lepidoptera: Pyralidae) was found to be the most potential predator on the wooly aphid. Activity of this predator was noticed in almost all sugarcane woolly aphid infested areas of Karnataka and Maharashtra. On an average, in Mandya during October, 8 larvae/leaf were found (n= 40 leaves). The larvae fed by constructing web and by remaining inside the web on ventral leaf surface. The movement of larvae from one leaf to another or from plant to plant was observed by hanging through silken thread. The feeding potential of larvae, on an average was 35-40 aphids/day (n=40). Pupation occurred on the undersurface of leaf and also in the leaf sheath. Adult females laid eggs on ventral surface of leaf among the colonies of SWA either singly or in a row. Among the five instars, fourth instar larvae were found to be efficient in feeding on a large number of aphids. Field collected larvae were reared in laboratory at 25-28° C and 70- 80% R.H. in Mandya. Adults are strong fliers with females laying eggs along the leaf midrib on both the sides. Each female deposited on an average 86±0.5 eggs (n=30). Neonate larvae ware greyish-white with brown head. The body was covered with white hairs. Grown up caterpillar was dark green with greyish brown head having longitudinal white streaks on the body.

The caterpillar constructed straw colored silken webbings all along the midrib and remained inside by feeding on the aphids. The early instar was inactive while the later was very active and voraciously fed on both aphid nymphs and adults. The larval period varied, an average of 20 to 35 days (n=28). The full grown larvae pupated inside the straw colored silken cocoons. The pupae were cylindrical, reddish with mean length of 8.9 mm (n=10). The pupal period ranged from 7 to 9 days (n=200). Adult longevity varied from 4 to 6 days (n= 20). The total life cycle varied from 30 to 50 days (n=180).

Larvae spun the silken webs and fed on the aphid during night. They remained inside the webbings during the day. Feeding potential increased with advancement of larval age. Under favorable climatic conditions 16 to 20 larvae per leaf were found during September-October 2004 in Mandya. In Dharwad, Shekarapppa (2003) recorded a maximum of 6-8 larvae/leaf under similar conditions.

Application of Bio-met, *Metarhizium anisopliae* @ 2ml/ litre and Biovert @ 3ml/litre of water and *Verticilium* @ 4 ml/ litre of water had no significant differences (CD at 5% by F test = 5.43 NS) in the number of leaves covered by aphids or leaves infested with aphids. The average number of aphid infested leaves under the above treatments were 6.85, 2.61, 3.4 respectively and in control the corresponding figure was 6.6%. Thus fungal pathogens accounted for very low mortality of woolly aphids. Even under field conditions the reduction in number of aphid infested cane leaves due to fungal pathogens was negligible.

Monitoring SWA infestation and number of predators began at both the localities viz., Mandya & Chikmagalur (Figs 2&3) when canes were over three months old and by August, the soils at base of most clumps were black with sooty mould fungus. Young tillers showed signs of drooping and wilting. By October canes and leaves drastically reduced in size and number. The aphid population inflicted severe damage to the crop. As a result, the farmers at Tarikere recorded 48% yield loss (Fig.2) and at Ghandal, Mandya, 43% (Fig.3). Field observations during 2002-2004 in South Karnataka revealed farmers incurring >50% losses in yields even with 20-25 predators/clump, on an average. Cane variety co. 8371 exhibited tolerance to SWA infestation compared to co. 86032. Patil *et al.,* (2004) recommended releases of predators especially *Dipha* and *Micromus* @ one larva/ 0.3 m^2 area. In the present study, on an average almost similar numbers of predators were obtained but on 5 clumps basis. Cultural and mechanical management tools must be adopted to add to the impact of predators in keeping SWA infestation below economic injury level. Thus integration of compatible management tools is warranted.

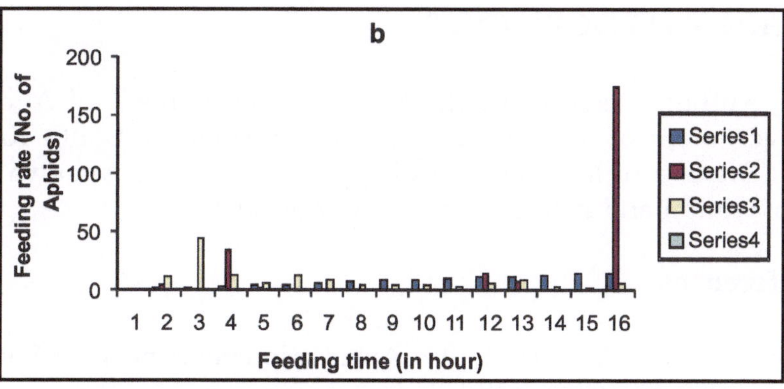

Fig 1. Feeding pattern of *Micromus* spp. in laboratory.

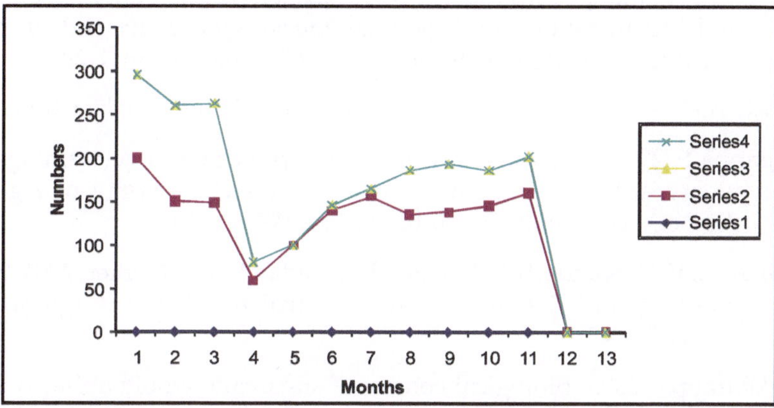

Fig 2. Seasonal occurrence of woolly aphids and their predators (Tarikere, Chikmagalur); Series .1 mean no. of infested leaves (n=25 clumps).2. mean no. of predators/25 clumps on co. 8371

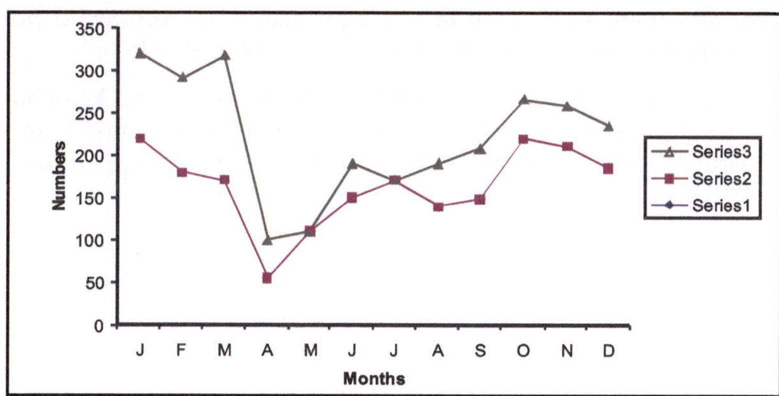

Fig 3. Mean occurrence of woolly aphids and their predators (Mandya), Series .1. mean no. of infested leaves (n=25 clumps).2. mean no. of predators/25 clumps. On co. 8603

ACKNOWLEDGEMENTS

Authors wish to thank officers of Department of Agriculture, Government of Karnataka, Sugarcane farmers in South Karnataka and Director of Instruction, College of Agriculture, Mandya for their cooperation and help in conducting this study.

References

Basu, A. N. & S. N. Banerjee. 1958. Aphids of economic plants of West Bengal. *Indian Agriculture*. 2:89-112.

David & V. Nandagopal.1986. Pests of Sugarcane, Distribution, Symptomology and Identification. In: *Sugarcane Entomology in India*. (Ed.) David *et al.*, Sugarcane Breeding Institute (ICAR), Coimbatore. pp. 1-29.

Ghosh, A.K. 1974. Aphids of economic importance in India. *Indian Agric.* 18(2):135.

Lingappa. S. 2003. World Scenario of Wooly aphid. State level training course on wooly aphid management in sugarcane. University of Agricultural Sciences, Dharwad. December 19-23, 2003. pp.1-6.

Patil, A. S., V. D. Shinde, S. B. Magar, N. G. Yadav & Y. S. Nerker. 2004. Sugarcane wooly aphid, its history and control measures. *Cooperative sugar* 36 (1): 37- 47.

Shekharappa. 2003. Biological control of sugarcane wooly aphid with special reference to *Dipha aphidivora* (Meyr.). State level training course on wooly aphid management in sugarcane. University of Agricultural Sciences, Dharwad. December 19-23, 2003. pp.17-18.

Vidya Mulimani, K. K. Patil & S. Lingappa. 2004. Coccinellid predators of sugarcane woolly aphid. *Insect Environ.*10 (1): 38-39.

Vidya Mulimani, G. K. Ramegowda.& M. S. Puttannavar. 2003. Potential predators of SWA, *Ceratovacuna lanigera* Zehnter. State level training course on wooly aphid management in sugar cane. University of Agricultural Sciences, Dharwad. December 19-23, 2003. pp.14-15.

41

Magnitude of Food Preference and Consumption by Soft-furred Field Rat, *Millardia meltada* (Gray) Under Laboratory Conditions

SAKTHIVEL, P. AND P. NEELANARAYANAN

ABSTRACT

Selection of acceptable bait is one of the most important elements for a successful rodent control campaign. Experiments were conducted by using the soft-furred field rat, *Millardia meltada* in order to find out its food preference between two forms *viz.*, germinated and non-germinated. Three cereals *viz.*, paddy (*Oryza sativa*), pearl millet (*Pennisetum typhoides*) and ragi (*Eleusine coracana*); and three pulses *viz.*, green gram (*Phaseolus aureus*), black gram (*Phaseolus mungo*) and Bengal gram (*Cicer arietinum*) were used in two forms during the present study. The animals' food preferences for six germinated grains were also found out. The results of the bi-choice tests revealed that both germinated cereals and pulses were preferred by the test animals of both sexes over the non- geminated ones. The differences between the quantities of consumption were statistically significant ($p<0.05$). The results of the acute toxicity tests (1.5 % Zinc phosphide) revealed that both sexes of metads attained 100% mortality under both no-choice and bi-choice (germinated paddy with poison vs plain non-geminated paddy bait) tests. The mean hours to death were 16.6 ± 7.89 (choice tests) and 5.4 ± 3.5 (no-choice tests). From the results of the present study it is suggested that the geminated cereals/pulses can be used as a poison bait carrier in general and either germinated green gram or germinated paddy in particular for the management of *M. meltada.*

INTRODUCTION

Rodents are a great menace to mankind as they share his food, damage his property and act as reservoirs of several dreadful diseases. They are known to be serious pests in farms, food stores and residential houses (Agrawal, 2000). They are versatile in feeding behaviour and choice of food. The selection of an acceptable bait or bait material is perhaps one of the most important elements for a successful rodent control campaign. An ideal rodenticide bait should be initially able to attract the rodent. Once the bait is attracted the target rodent species must be induced to eat the bait by virtue of its enhanced flavour, taste, palatability when compared to other available food (Sridhara & Srihari, 1986).

In view of the above, an attempt was made to study the food preference and consumption of germinated and non-germinated cereals (paddy, pearl millet and ragi) and pulses (green gram, black gram and Bengal gram) under bi-choice tests, individually. The Indian soft-furred field rat, *Millardia meltada,* one of the predominant rodent pests in *Cauvery* delta was subjected to this study with the following objectives:

1. To quantify the food preference and ingestion of germinated and non- germinated cereals (paddy, pear millet and ragi) and pulses (green gram, black gram and Bengal gram) (six tests individually) by the soft-furred field rat, *M. meltada,* under bi-choice tests.
2. To quantify the food preference and intake among the selected six germinated cereals and pulses under multiple-choice tests.
3. To study and record the mortality rate by using the most preferred germinated bait base at 1.5% concentration of Zinc phosphide under no-choice and choice tests.
4. To determine and suggest the most preferred poison bait carrier for future rodent pest management programmes.

MATERIAL & METHODS

Soft-furred field rats, *Millardia meltada* were live trapped from the crop fields near Mannachanallur, Tiruchirappalli District, Tamil Nadu by burrow digging method as suggested by Sivaprakasam (1988). They were acclimatized to the laboratory conditions for 10 days and during this period *ad lib* quantities of food and water was made available to them. Each test was conducted by using ten individually caged healthy animals (5 males + 5 females) after allowing them to starve for 24 hours.

The selected cereals, paddy (*Oryza sativa*), pearl millet (*Pennisetum typhoides*) and ragi (*Eleusine coracana*) and pulses, green gram (*Phaseolus aureus*), black gram (*Phaseolus mungo*) and Bengal gram (*Cicer arietinum*) were tested in two forms *viz.*, germinated and non-germinated.

FEEDING TRIALS

Bi-Choice Test: (Germinated *vs.* Non - Germinated): The individually caged rats were offered 20 g each of selected germinated and non-germinated cereals and pulses in separate containers.

Multiple-Choice Tests: (Germinated grains only): Twenty grams each of the selected germinated grains (cereals and pulses) were provided in separate containers to the individually housed *M. meltada*.

Both the bi-choice and multiple-choice tests were conducted for five consecutive days. The daily food consumption by the caged rodents was recorded for each individual for every 24 hours. The actual consumption was measured by using a Roy electronic balance with 0.05 g accuracy. The food in the respective forms and combinations were replenished once in 24 hours. All the values were converted to g/100g body weight of experimental animals. Water was provided *ad libitum* during the tests.

ACUTE TOXICITY TESTS

Zinc phosphide was used at 1.5% concentration as suggested by Neelanarayanan & Kanakasabai (1992). These tests were conducted under both no-choice and choice conditions. The most preferred grain, germinated paddy was used as the bait material for these tests.

No-Choice Tests: The poison bait was prepared with the following proportions (96.5 g germinated paddy + 1.5 g zinc phosphide + 2 g coconut oil). Each caged rat was offered a cup of 20 g poison bait and a cup of water. At the cessation of 24 hours, poison bait consumption by *M. meltada* was recorded. Further, the percentage of mortality and hours to death were also recorded.

Choice Tests: The poison bait was prepared as described above. Simultaneously, plain non-germinated paddy bait was also prepared (98 g plain non-germinated paddy + 2 g coconut oil). Each caged rat was offered a cup of 20 g poison bait, a cup of 20 g of plain bait and a cup of water.

After 24 hours, poison bait and plain bait consumption by *M. meltada* was recorded. Further, the percentage of mortality and hours to death were also recorded.

Student's 't' test was used in order to find out the differences between the germinated and non-germinated cereals and pulses consumption.

RESULTS & DISCUSSION

Bi-Choice Tests

Cereals: The mean daily consumption of germinated and non-germinated cereals (paddy, pearl millet and ragi) by *M. meltada* is presented in Table 1. Both sexes of the soft-furred field rat, preferred germinated cereals (Table 1). The most preferred food by both sexes of metads was ragi (8.45 ± 2.38 g), followed by pearl millet (5.94 ± 2.89 g) and paddy (5.28 ± 2.68g). The differences between the germinated and non- germinated cereal consumption was found to be statistically significant ($p < 0.05$). The percent increase in consumption of germinated paddy, pearl millet and ragi by both sexes of *M. meltada* were found to be 42.86 ± 37.49, 81.2 ± 27.16 and 79.59 ± 19.97, respectively.

Pulses: The mean daily ingestion of germinated and non-germinated pulses (green gram, black gram and Bengal gram) by *M. meltada* is presented in Table 2. From the results it is apparent that the germinated pulses were consumed in greater quantities than the non-germinated ones. The germinated green gram, black gram and Bengal gram were consumed at the rate of 9.78 ± 3.29 g, 9.12 ± 3.59 g and 9.79 ± 3.50 g, respectively by both sexes of metads. The maximum percentage of increase in consumption was noticed in Bengal gram (94.9 ± 10.39). The differences between the germinated and non-germinated pulse consumption by both sexes of metads was found to be statistically significant ($p < 0.05$). The percent increase in consumption of germinated green gram, black gram and Bengal gram over non-germinated forms by both sexes of *M. meltada* were found to be 91.31 ± 32.80, 79.09 ± 21.27 and 94.9 ± 10.39, respectively. The results of the present study are consistent with previous studies (Parshad & Nijjar, 1995; Vanitha *et al.*, 1997) in which the germinated grains were found to be preferred over non-germinated grains by *B. bengalensis*. The soft - furred field rats preferred the tested germinated grains in the following order, *i.e,* Bengal gram > green gram > pearl millet > ragi > black gram > paddy irrespective of the sexes under bi-choice tests (Tables 1 & 2).

Multiple-choice tests

The mean daily ingestion of germinated cereals and pulses by *M. meltada* is presented in Table 3. In multiple-choice tests, the germinated grains consumed by both sexes of the soft furred field rats were found to be in the preferential order of green gram (4 ± 3 g or 11.4 ± 8.1 g/100 g body weight) > paddy (3 ± 2 g or 7.7 ± 5.4 g/100 g body weight) > pearl millet (2 ± 2 g or 6 ± 4.7 g/100 g body weight), = bengal gram (2 ± 2 g or 6.5 ± 5.5 g/100 g body weight) > ragi (1 ± 1 g or 4.1 ± 3.1 g/100 g body weight) > black gram (1 ± 0 g or 2.4 ± 1.4 g/100 g body weight) (Table 3). According to Renapurkar & Deoras (1985), the selection of acceptable bait or bait material is perhaps the most important element for a successful rodent control campaign. The ideal rodenticidal bait should be initially able to attract the rodent. Once the bait is attracted the target rodent species must be induced to eat the bait by virtue of its enhanced flavour, taste, palatability when compared to other available food (Sridhara & Srihari, 1986).

ACUTE TOXICITY STUDIES

In the present study, the rodents preferred the germinated cereals and pulses over the non-germinated ones. Among them, germinated paddy was used as a principal bait material. The mortality rate of *M. meltada* after being fed by 1.5% zinc phosphide under no-choice and choice tests is provided in Table 4.

Table 1. Mean daily intake of germinated and non-germinated cereals by *M.meltada*

Sl.No. & Name of the cereals	sex	Body weight of the Experimental Animals (g) X ± SD	Germinated (g) X ± SD	Non-Germinated (g) X ± SD	Amount of Cereals Consumed: % increase in consumption X ± SD	Germinated (g/100g body wt.) X ± SD	Non-Germinated (g/100 g body wt) X ± SD	Total Consumption (Germinated + Non-Germinated) (G/100 g body wt.) X ± SD
1. Paddy	Male*	47.6 ± 2.24 (45.50)	6.78 ± 1.79* (2.3-9.2)	1.74 ± 1.55 (0.1-14.27)	72.22 ± 21.08 (21.21-99.27)	12.66 ± 4.43 (7.3-19.4)	4.28 ± 3.80 (0.05-6.85)	18.88 ± 5.13 (7.66-28.3)
	Female*	64.2 ±20.7 (36-95)	5.28 ± 2.68* (0.4-6.95)	3.68 ± 1.93 (1.8-13.72)	42.86 ± 37.49 (0.87-97.83)	8.94 ± 4.38 (0.72-19.3)	6.65 ± 4.00 (0.2-7.55)	15.59 ± 5.94 (8.52-28.7)
Total	⚥	55.9 ± 16.96 (36-95)	5.28 ± 2.68 (0.4-6.95)	3.68 ± 1.93 (1.8-13.72)	42.86 ± 37.49 (0.87-97.83)	8.94 ± 4.38 (0.72-19.3)	6.65 ± 4.00 (0.2-7.55)	15.59 ± 5.94 (8.52-28.7)
2. Pearl Millet	Male*	47.6 ± 2.19 (45-50)	5.17 ± 1.94* (2.25-11.35)	0.93 ± 1.88 (0-3.75)	79.21 ± 29.35 (8.45-100)	11.38 ± 3.81 (5-23.16)	1.98 ± 2.58 (0-8.35)	12.80 ± 3.60 (7.32-23.16)
	Female*	63.6 ± 21.11 (36-94)	6.61 ± 3.53* (1.4-15.55)	1.77 ± 2.77 (0-12.35)	83.18 ± 25.09 (31.03-100)	10.91 ± 5.84 (2.59-27.36)	3.48 ± 5.54 (0-22.87)	14.39 ± 6.12 (6.43-27.36)
Total	⚥	5.50 ± 16.74 (36-94)	5.94 ± 2.89 (1.4-15.55)	1.35 ± 2.13 (0-12.35)	81.2 ± 27.16 (8.45-100)	10.86 ± 4.82 (2.59-27.36)	2.73 ± 4.30 (0-22.87)	13.60 ± 4.98 (6.43-27.36)
3. Ragi	Male*	4.38 ± 5.18 (38-51)	8.30 ±2.37* (2.05-12.65)	1.54 ± 1.28 (0-3.65)	82.98 ± 14.97 (54.93-100)	19.20 ± 5.80 (4.01-33.40)	3.56 ± 3.05 (0-9.21)	22.77 ± 5.78 (8.51-34.86)
	Female*	65.8 ± 13.75 (54-86)	8.58 ± 2.48* (3.35-12.65)	2.10 ± 2.47 (0.05-12.05)	75.35 ± 24.15 (34.42-100)	13.60 ± 4.58 (3.89-23.42)	2.69 ± 4.14 (0.05-21.51)	16.94 ± 6.06 (6.85-35.34)
Total	⚥	54.8 ± 14.99 (38-86)	8.45 ± 2.38 (2.05-12.65)	1.82 ± 1.95 (0-12.05)	79.59 ± 19.97 (34.42-100)	16.40 ± 5.84 (3.89-33.42)	3.45 ± 3.66 (0-21.51)	19.76 ± 7.25 (6.85-35.34)

* Mean and standard deviation values were from twenty-five individual observations

· The difference between the germinated and non-germinated cereal consumption was statistically significant ($P < 0.05$) (Student's " t " test)
Figures within the parentheses indicate the range value

Table 2. Mean daily intake of germinated and non-germinated pulses by *M. meltada*

Sl.No. & Name of the pulses	Sex	Body weight of the Experimental Animals (g) X ± SD	Germinated (g) X ± SD	Non-Germinated (g) X ± SD	% increase in consumption X ± SD	Amount of pulses consumed: Germinated (g/100g body weight) X ± SD	Amount of pulses consumed: Non-Germinated (g/100 g body weight) X ± SD	Total Consu - mption (Germinated+Non Germinated) g/ 100 g body wt.)
	Male *	47.6 ± 2.15 (45-50)	10.78 ±2.41[?] (6.4-18.70)	1.04 ± 1.14 (0.2-3.20)	89.34±11.49 (74.19-100)	22.59±4.70 (13.06-29.21)	2.25 ± 2.47 (0.2-6.66)	24.85 ± 5.27 (13.06-32.66)
1.Green gram	Female*	63.23 ± 20.36 (36-94)	8.73 ± 3.81[?] (0.5-15.75)	1.24 ± 1.07 (0.4-3.65)	81.19± 20.82 (27.72-100)	14.33 ± 6.03 (0.53-25.2)	2.31± 2.68 (0.42-10.13)	16.67 ± 6.91 (1.16-26.85)
Total	⚥	55.6 ± 16.74 (36-94)	9.78 ± 3.29 (0.5-18.70)	1.78 ± 2.12 (0.2-3.65)	91.35 ± 32.80 (27.72-100)	18.46 ± 6.81 (0.53-29.21)	3.01 ± 2.57 (0.2-10.13)	20.76 ± 7.38 (1.16-32.66)
	Male *	54.8 ± 12.46 (41-78)	8.41 ± 3.90[?] (1.7-12.9)	1.88 ± 1.84 (0-7.85)	74.84 ± 23.20 (22.10-100)	15.83 ± 8.54 (4.14-36.95)	3.64 ± 3.73 (0-14.53)	19.48 ± 9.94 (6.7-48.79)
2.Black gram	Female*	43.2 ± 4.71 (38-50)	9.84 ± 3.27[?] (5.35-16.55)	1.81 ± 1.45 (0-4.35)	79.26 ± 19.95 (59.89-100)	23.07 ± 7.79 (11.88-35.39)	4.34 ± 3.60 (0-10.72)	27.42 ± 9.27 (13.6-38.67)
Total	⚥	49 ± 11.18 (38-78)	9.12 ± 3.59 (1.7-16.55)	1.84 ± 1.62 (0-7.85)	79.09 ± 21.27 (22.10-100)	19.45 ± 8.87 (4.14-36.95)	3.99 ± 3.60 (0-14.53)	23.45 ± 10.22 (6.7-48.79)
	Male*	48.83 ± 3.43 (45-50)	9.89 ± 3.43[?] (5.2-18.05)	0.40 ± 0.70 (0.15-2.85)	95.67 ± 7.47 (83.49-100)	20.84 ± 6.87 (7.9-37.6)	1.24 ± 2.35 (0.33-10.4)	22.65 ± 8.46 (7.9-48)
3. Bengal gram	Female*	64.2 ± 20.79 (36-95)	9.92 ± 3.57[?] (2.2-15.4)	0.46 ± 0.99 (0.05-3.85)	91.06 ± 21.24 (55.49-100)	15.67 ± 6.12 (6.05-27.9)	1.8 ± 4.08 (0.05-16.6)	67.52 ± 7.12 (0.05-34.71)
Total	⚥	55.9 ± 16.96 (36-95)	9.79 ± 3.50 (2.2-18.05)	0.44 ± 0.86 (0.05-3.85)	94.9 ± 10.39 (55.49-100)	18.25 ± 7.0 (6.05-37.6)	1.58 ± 3.32 (0.05-16.6)	20.09 ± 8.23 (0.05-48)

* Mean and standard deviation values were from twenty-five individual observations

• The difference between the germinated and non-germinated pulse consumption was statistically significant ($P < 0.05$) (Student's " t " test); Figures within the parentheses indicate range values.

Table 3. Mean intake of germinated cereals and pulses by M. *meltada* under multiple choice tests

FOOD	Daily intake (g) X ± SD[1]			Intake of grains (g/100g body wt.) X ± SD[1]		
	Male (41 ± 11) *	**Female (36 ± 2) ***	**⚥ (39 ± 8) ***	**Male (41 ± 11) ***	**Female (36 ± 2) ***	**⚥ (39 ± 8) ***
Paddy	4 ± 2 (0.5-8.55)	2 ± 1 (0.35 - 3.85)	3 ± 2 (0.35 - 8.55)	7.7 ± 6.2 (0.8 - 21.9)	5.7 ± 3.3 (1.7 -11.3)	7.7 ± 5.4 (0.8 - 21.9)
Pearl millet	2 ± 2 (0.55-9)	3 ± 2 (0.9-6.6)	2 ± 2 (0.55-9)	4.2 ± 3.6 (1.4 - 11.6)	7.9 ± 4.8 (3.1-19.4)	6 ± 4.7 (1.4 - 19.4)
Ragi	2 ± 1 (0.6 - 4.05)	1 ± 1 (0.7 - 3.6)	1 ± 1 (0.6 - 4.05)	4.3 ± 3.5 (1.2 - 14.5)	3.8 ± 2.7 (1.6 -10.6)	4.1 ± 3.1 (1.2 - 14.5)
Green gram	2 ± 2 (0.5 - 7.4)	6 ± 3 (1.6 - 9.55)	4 ± 3 (0.5 - 9.55)	6.3 ± 4.9 (1.2 - 18.5)	16.5 ± 7.4 (4.3 - 29)	11.4 ± 8.1 (1.2 - 29)
Black gram	1 ± 0 (0.3 - 1.35)	1 ± 1 (0.1 - 2.1)	1 ± 0 (0.1 - 2.1)	2.1 ± 0.9 (1.2 - 4.3)	2.8 ± 1.6 (0.3-6.2)	2.4 ± 1.4 (0.3 - 6.2)
Bengal gram	3 ± 2 (0.7 - 6.8)	2±1 (0.6-5.5)	2 ± 2 (0.6 - 6.8)	8.2 ± 6.6 (1.2 - 30.6)	4.7 ± 3.4 (1.3-16.2)	6.5 ± 5.5 (1.2 - 30.6)
Overall cons umption	13.53 ± 2.27 (10.3-19.55)	14.81 ± 3.13 (9.6 - 22.2)	13.65 ± 3.66 (9.6 - 22.2)	33.9 ± 8.2 (21 - 50.6)	41.4 ± 9.5 (25.9 - 5.3)	38.2 ± 9.6 (21 - 65.3)

* Mean ± SD body weight (g) of the experimental animals
Figures within the parentheses indicate range values.

1 Mean and standard deviation values were for twenty-five individual observations

Table 4: Mortality rate of *M. meltada* after being fed 1.5% zinc phosphide (germinated paddy as principal bait) under choice and no-choice tests

Parameters	Choice test ⚥			No-Choice test ⚥		
	Male[@] **X ± SD**	**Female**[@] **X ± SD**	**X ± SD**	**Male**[@] **X ± SD**	**Female**[@] **X ± SD**	**X ± SD**
Body weight (g)	42.4 ± 13.35 (25 – 60)	34.4 ± 2.60 (30 – 37)	38.4 ± 10 (25 – 60)	41 ± 10.4 (28 – 60)	36 ± 2.3 (34 – 40)	38.5 ± 8 (28 – 60)
Intake of poison bait in germinated bait (g)	2.56 ± 1.39 (1.5 – 4.94)	1.61 ± 0.44 (1.32 – 2.38)	2.09 ± 1.09 (1.32 – 4.94)	–	2 ± 1.1 (0.9-3.4)	1.9 ± 0.9 (0.9-3.4)
Intake of plain bait (g)	0.2 ± 0.2 (0.1 - 0.7)	0.1 ± 0.1 (0 – 0.8)	0.3 ± 0.3 (0 – 0.8)	1.8 ± 0.5 (1.2 - 2.5)	–	–
Intake of poison bait (g/100 g body weight)	6.26 ± 3.33 (3.76 – 12.04)	4.68 ± 1.09 (3.74 – 6.43)	.47 ± 2.48 (3.74 – 12.04)	4.3 ± 0.9 (3.1 – 4.3)	5.7 ± 3.2 (2.3 – 10)	5 ± 2.4 (2.3 – 10)
Intake of plain bait (g/100 g body weight)	0.6 ± 0.6 (1.2 – 1.8)	0.3 ± 0.3 (0 – 0.8)	0.5 ± 0.5 (0 – 1.8)	–	–	–
Mortality rate (%)	100%	100%	100%	100%	100%	100%
Hours to death	9.4 ± 7.73 (4 – 23)	23.8 ± 28.16 (8 – 23)	16.6 ± 20.89 (4 - 23)	6.6 ± 3.9 (3 – 14)	4.3 ± 2.4 (2.3 – 9)	5.4 ± 3.5 (2.3 – 14)

@ Mean and standard deviation values were from five individual observations
Figures within the parenthesis indicate range values.

Choice tests: The observed mean consumption of poison bait by males was 6.26 ± 3.33 g/100 g body weight with a range of 3.76–12.04 g /100 g body weight. The observed mean ingestion of poison bait by females was 4.68± 1.09g/100 g body weight with a range of 3.74 - 6.43 g/100 g body weight. Cent percent mortality of *M. meltada* was observed when fed on 1.5 % Zinc phosphide in bait under no-choice tests. Mortality of animals was noticed between 4-23 hrs in both sexes with mean of 9.4 ± 7.73 for males and 23.8 ± 28.16 for females.

No-choice tests: The observed mean consumption of poison bait by males was 4.3 ± 0.9 g/100 g body weight with a range of 3.1–4.3 g / 100 g body weight. The observed mean ingestion of poison bait by females was 5.7± 3.2g/100 g body weight with a range of 2.3–10 g/ 100 g body weight. Cent percent mortality of *M. meltada* was observed on feeding with 1.5 % zinc phosphide. Mortality of animals was observed between 2.3–14 hrs in both sexes with mean of 6.6 ± 3.9 (males) and 4.3 ± 2.4 (females). Contrary to this, Neelanarayanan & Kanakasabai (1992) reported that 50% mortality was observed among *M. meldata* after feeding with 2% zinc phosphide. The results clearly revealed that the poison when mixed with germinated grains has induced *M. meldata* to consume greater quantities of bait. This may be due to the poison carrier's i.e germinated seed's enhanced palatability and texture.

A series of bi-choice tests and multiple choice tests carried out in the present study indicated that the germinated cereals and pulses can be suggested as a poison carrier for rodent control programme in general and either germinated green gram or germinated paddy in particular for the management of *M. meltada*. Therefore, the germinated paddy can be substituted over other forms of rice such as, whole grain, flour, cracked and roasted as a bait carrier and the same can be suggested to the farmers for getting better results in rodent control campaigns. However, this requires further field trails for a final recommendation.

ACKNOWLEDGEMENTS

The financial assistance rendered by UGC, Hyderabad (Link No.: 1804 February 2005) for carrying out this research work is gratefully acknowledged. The authors are thankful to Dr. S. Ramalingam, Principal and Management of Nehru Memorial College (Autonomous) for having rendered necessary facilities. We are also thankful to the rodent trappers who helped us to trap the rodents for these experiments. The moral support and encouragement extended by Prof. K. Kanagasundari (Zoology), Nehru Memorial College (Autonomous), Puthanampatti are gratefully acknowledged.

References

Agrawal, V.C. 2000. Taxonomic Studies on Indian Muridae and Hystricidae (Mammalia: Rodendia). *Rec. Zool. Surv. India,* Occasional Paper No. 180. i-viii, 1- 186 p.

Neelanarayanan, P. & R. Kanakasabai. 1992. Laboratory evaluation of zinc phosphide and its impact on biochemistry of the soft furred field rat, *Millardia meltada* (Gray). *J.Ecotoxicol.Environ. Monit.* 2(4): 273-277.

Parshad, V.R. & C. Nijjar. 1995. Effects of Rice variety and germination on the feeding preferences of the lesser Bandicoot rat, *Bandicota bengalensis. Rodent Newsl.* 19 (2): 6 – 7.

Renapurkar, D. M. & P. J. Deoras. 1985. Some observations on food habits of *Rattus rattus,* the house rat. *Pestology* 9(6): 41-44.

Sivaprakasam, C.1988. *Studies on the burrow ecology, population and food habits of field rodents in and around Manampandal and Arupathy areas.* M.Sc. Thesis, Bharathidasan University, Tiruchirappalli, Tamil Nadu.

Sridhara, S. & K.Srihari.1986. Bait preference of rodent pests of Karnataka - *Bandicota bengalensis, Tatera indica, Rattus meltada* and *Mus booduga. Mysore J. agric.Sci.* 20: 296 – 300.

Vanitha,V. P. Neelanarayanan & R. Kanakasabai. 1997. Laboratory evaluation of germinated cereal as a bait base for the control of Indian mole rat, *Bandicota bengalensis* (Gray). *Pestology* 31(6): 38-42.

42

Oriental Responses of Shoot and Fruit Borer Larvae, *Conogethes punctiferalis* Guenee (Pyralidae: Lepidoptera) Towards Selected Plants in Laboratory

CHAKRAVARTHY A.K.
B.S. BASAVARAJU AND E. GANGAPPA

ABSTRACT

The shoot and fruit borer, *Conogethes punctiferalis* is a serious pest on several cultivated plants and the population consists of two types, viz., stalk feeding type (SFT) and fruit feeding type (FFT). The SFT and FFT *Conogethes* larvae perceived their host plants from 1.2 to 2.0 cm distance depending on the age of the larva. The minimum distance for the host perception increased with the increase in the age and size of the larva.

All the instars of both the types responded and orientated readily to the host stimulus. On reacing the host, larvae palpated and bored in to the shoot/fruit. Orientation consisted of random and directed movements accompanied with raising and lowering the head. FFT larvae preferred Castor and SFT, cardamom. The inactivation of sense organs involved in host perception was accomplished by cold (-5° C), hot (35° C) and acid treatment. Experiments indicated that relative suitability of plants as hosts for *Conogethes* larvae types depended on differences in the responses involved in the six different phases of their establishment on the plant. These observations have implications for managing *Conogethus* on different cultivated crops/plants.

INTRODUCTION

Cardamom (*Elettaria cardamomum* Maton) is an important spice crop popularly known as "Queen of Spices". In India, cardamom is cultivated in 84,000 ha. Kerala accounts for 60 per cent with a production of 5050 metric tonnes followed by Karnataka and Tamil Nadu accounting for 30 and 10 per cent area with a production of 1700 and 750 metric tonnes, respectively (Anon, 1996). Many workers have studied the biology of *Conogethus punctiferalis* Guanee that attacks many crops in different parts of the world. Krishnamurthy *et al.*, (1989) reported that crop loss due to shoot and fruit borer varied from 0.92 to 10.00 per cent. Similar report on the extent of crop loss was reported by Anon (1997). Kapadia (1996) estimated 42.3 per cent crop loss in castor in India. Fifty per cent yield reduction was recorded on grapes due to *Conogethus punctiferalis* attack (Ram *et al.*, 1997). Thyagaraj (2002) developed integrated management practices for *C. punctiferalis* on cardamom in Karnataka. In this paper, field and laboratory observations on orientational behavior of *C. punctiferalis* are reported.

MATERIAL & METHODS

Based on the published literature, the available host plants in the region were selected and tested in the laboratory. Separate sets of petriplates (10" dia) were maintained in which tender plant parts of (<1 year old) different host plants were exposed to the freshly emerged *Conogethus* larvae collected from the culture maintained in the laboratory. The neonate larvae were released at the centre of the plastic box (12" dia) where the fresh plant part pieces (2.5 cm length) were placed on a spread of moist cotton covered with filter paper of same size on the top. The box was placed in a wire mesh insect cage (0.3 m^3) with zero watt red light bulb on the top and covered with black paper on all other sides. The larvae were starved for 30 minutes before release. The movements of the larvae were monitored for 15 minutes. The larvae were released at different distances (Table.1) from the stimulus source. If the larvae made direct and rapid movement towards the stimulus it was designated as taxis. If the larvae made random movements, the behavior was designated as kinesis. Based on the movement and behaviour, the responses of larvae to plant extracts were categorized as attractant, repellent, neutral and arrestant. Attractant if larvae oriented to the stimulus, Repellent if the larvae moved away from the stimulus, Arrestant if the larvae were rendered immobile and Neutral if larvae made random movements.

Table.1. Effective distance for perception of stimulus source in SFT and FFT *Conogethus* larvae.

Conogethus	**Larval instar type larvae**	**Distance from the stimulus source (cm)**	**Behavioral responses**
SFT	Fourth	3.0	Random movements, Intermittently raising anterior part of the body, legs and head.
SFT	Fourth	1.5	– – –Do– – –
SFT	Fourth	1.2	Within 25 sec.of release, larvae exhibited directed movements.
SFT	Third	1.5	Exhibited kinesis, Intermittently raising anterior part of the body, legs and head
SFT	Third	1.2	Exhibited taxis; reached stimulus within 20 sec of release
SFT	First	0.32 to 0.5	Showed taxis and palpated on the stimulus within 1 sec. of release
FFT	First	0.22 to 0.3	– – – –Do– – – –
FFT	Second	0.5 to 0.8	– – – –Do– – – –. The larvae perceived the host within 30sec. of release
FFT	Third & Fourth	1.5 to 2.0	– – – –Do– – – –

SFT= stalk feeding type: FFT= Fruit feeding type; larvae were starved for atleast 30 min. before the tests; fresh larvae were used for every test.

Various sense organs of the first instar *Conogethus* larvae were given heat treatment to antennae, maxillae and others as indicated in Table 2. A soldering gun clamped with iron wire was held over the body structure of larvae for about a minute and the larvae so treated were released at the center of the plastic container and orientation behavior recorded as mentioned above. Similarly orientation responses of first instar FFT and SFT larvae were determined to volatile extracts of selected plant species. A known weight of plant part (10g) was crushed in 250ml of the test solvent. The extract was filtered using a funnel and filter paper and the filtrate was dried, concentrated and kept under a deep freezer for experimental purpose.

Table 2. Orientation responses of *Conogethus* larvae to host plants after heat treatment of various sense organs.

Larval sense organ	Orientational responses of larvae		
	SFT	FFT	REMARKS
Antennae	_	_	Larvae do not orient
Maxillae	_	_	Larvae made random movements
Maxillary palps	_	_	Larvae do not perceive host from 5 mm
Labium	_	_	Larvae exhibited random movements after 1 hr
Labial palps	_	_	-do-
Hairs on head	+	+	Larvae perceived host within 5.0 mm
Mouth parts	_	_	Larvae do not orient even from 5.00 mm
Cervix	+	+	Perceived host within 30 sec
First pair of tarsi	+	+	Perceived host within 100 sec.
Second pair of tarsi	+	+	-do-
Third pair of tarsi	+	+	-do-
All three pairs of tarsi	_	_	Larvae remained immobile. So unable to orient.

+ = Orientational response; ¯ = No. Orientation responses of larvae for 6 hr. after release.

Chemical grade methanol, acetone, sodium hydroxide and ether were used as solvents. The first , second and third instars larvae of *Conogethus* were also used to inactivate sense organs (see Table 2) by treating with concentrated hydrochloric acid. The tip of needle was dipped in hydrochloric acid and held over the sensory structure of the *Conogethus* larva for a minute and the responses of larvae to different plants determined. Both SFT and FFT second/ third instar larvae were starved for 6 to 8 hr. and maintained over a pack of ice for two minutes.

Cold treated larvae were then given heat-treatment using a soldering gun clamped with a thin wire. After heating to red- hot, the wire was placed over the part to be rendered defunctional for a minute. After heat treatment, larvae were maintained in petridish over moist cotton wad for 40 to 60 min. Subsequently larvae were placed at the centre of a petridish with respective host plants separately and observations on orientation responses recorded. The treatments, heat/cold/acid were repeated thrice to ensure inactivation of the concerned sensory structure.

RESULTS & DISCUSSION

The SFT and FFT *Conogethus* larvae perceived their host plants within a distance of 1.2 to 2.0 cm depending on the stage of the larva. The behavior exhibited by the larvae to the host stimulus is summarized in Table 1. The minimum distance for the host perception increased with the increase in the age and size of the larvae in both the *Conogethus* types.

The first instar larvae of both the *Conogethus* types responded and orientated readily to the host stimulus. On arrival at the host, larvae palpated and bored into the shoot/ fruit. When the starved first instar SFT larvae were released 0.5 cm away from castor, the larvae (n=10) did not orient towards the plant stimulus source and were repelled instead. However, when starved first instar FFT larvae were released 0.5 cm away from cardamom stalk, the larvae randomly neared the source, moved extensively over the cardamom stalk and then moved away from the source.

Close range orientational responses of starved, fourth instar SFT larvae oposed to castor and cardamom in an insect cage ($0.3m^3$), showed that larvae (n=12) moved away from castor fruits but orientated to the cardamom in 50 sec. When castor fruits were arranged radially inside the cage and SFT larvae released 1.5cm away from each fruit, larvae made random movements, nibbled at the fruit and then moved away towards centre. When starved FFT third instar larvae were released 2.0 cm away from cardamom stalk, the larvae after making undirected movements initially, landed on the stalk, perceived the stalk by touch and initiated bites, then moved away (n=8).

Orientation responses of second instar SFT larvae to castor showed that the larvae made random movements, nibbled at the fruits and moved towards the walls of the cage. When second instar FFT larvae were released to cardamom stalk at 2.0 cm distance, larvae exhibited directed movements towards the stalk in 55 seconds without palpating at the stalk. These observations suggested that SFT and FFT *Conogethus* larvae exhibited preferences for plants different from one another. However, the larvae of both the types are adaptable to the same plant species.

When various sense organs of *Conogethus* larvae were heat treated, they got disorientated, made random movements or took time to perceive the host stimulus (Table 2). When antennae and maxillae were inactivated larvae disorientated. When maxillary palps and three pairs of tarsi were inactivated, larvae did not perceive the host stimulus. These observations highlight the roles of different sense organs in the perception and recognition of the host plant.

When the mouth parts of *Conogethus* larvae were inactivated by acid, larvae could perceive the stimulus only after 3.5 hr from 30 cm, but could not bite the plant part as mandibles were injured. When maxillae, antennae and labial complex were acid treated, larvae after 50 min. orientated a bit and consumed the plant part. So it can be concluded that acid temporarily impaired the sensory structures.

Results of orientation responses of first instar *Conogethus* larvae to three solvents of cardamom showed that ether, methanol and ethyl acetate extracts were attractive to larvae but not sodium hydroxide extract. *Amomum subulatum* extracts of acetone and ethyl acetate were not attractive to larvae. Water extracts of *Hedycium flawasense* were attractive to larvae but not of sodium hydroxide and ethyl acetate extract (Table 3).

Conogethus larvae were exposed to volatile extracts of different solvents and they responded differently. For instance when cardamom was used as a host the larvae were attracted to extracts of all the three solvents. The larvae were attracted to methanol extract when *Alpinia* was used as the plant source. The larvae responded to ether extract of *Hedycium caronarium* but not to methanol and acetone extracts. Interestingly larvae were also attracted to extracts of the three solvents when *Zingiber montanum* was used as the plant source (Table 4).

Similar trend in the orientation responses of first instar larvae of fruit feeding type to solvent extracts of plants was found. The details are depicted in Table 5. It was found that larvae responded to extracts eliciting positive stimuli and were repelled away when the extracts contained a repellent compound or lacked an attractant. Changes in feeding orientation responses may be attributed to the changes in receptors on the sensory structures (Bernays *et al.*, 1972).

Table 3. Orientational response of first instar stalk feeding type (SFT) *Conogethus* larvae to volatile extracts of different plants in different solvents.

Plant	Solvent extract	Responses of Larvae			
		Attractant	Repellent	Neutral	Arrestant
Alpinia neutani	Methanol	?– ??+	– –	+ –	– –
	Sodium hydroxide	+ +	– –	– –	– –
	Ethyl acetate	– –	– –	+ +	– –
	Acetone	+ +	– –	– –	– –
	Ether	– –	– –	+ +	– –
Hedycium coronarium	Water	+ +	– –	– –	– –
	Ethyl acetate	– –	– –	+ +	– –
	Sodium hydroxide	– –	+ +	– –	– –
Ellettaria cardamomum	Ether	+ +	– –	– –	– –
	Methanol	+ +	– –	– –	
	Ethyl acetate	+ +	– –	– +	– –
	Sodium hydroxide	– –	+ +	– –	– –
Ricinus commonis	Acetone	– –	– –	+ +	– –
	Ether	+ +	– –	– –	– –
	Ethyl acetate	– –	– –	+ +	– –
Elettaria cardamom (Semi-erect type)	Methanol	– –	– –	+ +	– –
	Ether	+ +	– –	– –	– –
	Ethyl acetate	– –	– –	+ +	– –
Hedycium sp	Ether	– –	+ +	– –	– –
	Water	– –	– –	+ +	– –
	Sodium hydroxide	– –	– –	+	– –

?= Replication1; ??= Replication 2.

Table 4. Orientational responses of fourth instar stalk- feeding type (SFT) *Conogethus* larvae to volatile extracts of different plants in different solvents.

Plant	Solvent extract	Response of larvae			
		Repellant	Attractant	Neutral	Arrestant
Alphinia neutan	Methanol	− −	+ ±	− −	− −
	Acetone	− −	−	+ ±	− −
	Ether	−	±	−	− −
Hedycium sp	Methanol	+ ±	− −	− ±	− −
	Acetone	+ +	− −	−	− −
	Ether	±	− +	− −	− −
Elettaria cardamom (Semi – erect)	Methanol	−	±	− −	− −
	Acetone	+ ±	− −	−	− −
	Ether	− −	+ +	+ ±	− −
Hedycium coronarium	Methanol	− −	±	− +	− −
	Acetone	− −	− −	+ ±	− −
	Ether	− −	+ +	−	− −
Eletttaria cardamomum (Erect type)	Methanol	− −	+ +	− −	− −
	Acetone	− −	±	− ±	− −
	Ether	− −	+ +	−	− −
Eletttaria cardamomum (creeping type)	Methanol	− −	− −	+ +	− −
	Acetone	− −	− −	+ ±	− −
	Ether	− −	+ +	−	− −
Colacasia species	Methanol	− −	+ ±	− −	− −
	Acetone	− −	−	+ ±	− −
	Ether	− −	±	+	− −
Ricinus commonis	Methanol	+ +	− −	− −	− −
	Acetone	+ ±	− −	− −	− −
	Ether	−	±	− +	−
Amomum subulatum	Methanol	±	− −	− ±	− −
	Acetone	+ ±	− −	− −	− −
	Ether	−	+ +	−	−
Zingiber montanum	Methanol	− −	+ +	− −	− −
	Acetone	− −	+ +	− −	− −
	Ether	− −	+ +	− −	− −

Table 5. Orientational responses of fourth instar Fruit feeding type (FFT) *Conogethus* larvae to volatile extracts of different plants in different solvents

Plant	Solvent	Response of larvae			
		Repellent	Attractant	Neutral	Arrestant
Alpinia neutan	Methanol	− −	− −	+ ±	− −
	Acetone	− −	+ +	−	− −
	Ether	− −	+ +	− −	− −
Hedycium sp	Methanol	− −	− −	+ ±	− −
	Acetone	±	−	−	− −
	Ether	− −	± −	− −	− −
Hedycium coranarium	Methanol	− −	±	−	− −
	Acetone	− −	±	±	− −
	Ether	− −	+ +	± −	− −
Elettaria cardamomum (semi-erect type)	Methanol	− −	+ ±	− −	− −
	Acetone	+ ±	−	− −	− −
	Ether	−	+ +	− −	− −
Elettaria cardamomom (Erect type)	Methanol	+ ±	− −	− −	− −
	Acetone	−	+ +	− −	− −
	Ether	− −	+ +	− −	− −
Elettaria cardamomom (Creeping type)	Methanol	− −	±	− +	− −
	Acetone	− −	+ +	+ +	− −
	Ether	− −	+ +	+ +	− −
Colocasia sp	Methanol	±	− +	− −	− −
	Acetone	− −	±	−	− −
	Ether	− −	±	± +	− −
Ricinus commonis	Methanol	− −	+ +	− −	− −
	Acetone	− −	+ +	− −	− −
	Ether	− −	+ +	− −	− −
Amomum subulatum	Methanol	− −	±	− +	− −
	Acetone	− −	− −	+ +	− −
	Ether	− −	− +	±	− −
Zingiber montanum	Methanol	− −	±	− ±	− −
	Acetone	− −	+ +	−	− −
	Ether	− −	±	− +	− −

ACKNOWLEDGEMENTS

Authors thank Dr. N.E. Thyagaraj, Associate Director of Research and staff of Regional Research Station, Mudigere for the help rendered during field investigations and data collection.

References

Anon 1996. *Annual Report*. Indian Institute of Spice Research, Calicut, Kerala. 23 pp.

Anon 1997. *Annual Report of AICRP on Spices*. University of Agricultural Sciences, Bangalore. pp.7-8.

Bernays, E. A., W. M. Blaney & R. F. Chapman. 1972. Changes in chemoreceptor sensilla on the maxillary palps of *Locusta migratoria* in relation to feeding. *J. exp. Boil.* 57: 745-753.

Kapadia, M.N. 1996. Estimation of losses due to pod borers in oil seed crops. *J. Oilseeds Res.* 13 (1): 139-140.

Krishnamurthy, K., M. M. Khan, K. K. Avadhani, J. Venkatesh, A. L. Siddaramaiah, A. K. Chakravarthy & B. R. Gurumurthy. 1989. *Three Decades of Cardamom Research.* (1958-1988). Technical Bulletin. Regional Research Station, University of Agricultural Sciences (Bangalore), Mudigere. pp.44-68.

Ram, K. T., P. S. Hugar, Somasekar & B. V. Patil. 1997. Record of *Conogethus punctiferalis* on grapes. *Pest Management in Horticultural Ecosystems* 3:37.

Thyagaraj, N.E. 2002. *Integrated Management of some important Cardamom (Elettaria cardamomum* Maton) *pests in Hill region of Karnataka, South India.* Ph.d. thesis submitted to Dr. B.R. Ambedkar University, Agra. 226 pp.

43

Foraging Behaviour of Honey Bee and Its Role in Fruit Set in Cucumber

BHAT, N.S., H.S.SATHISH AND H.M.YASHWANT

ABSTRACT

Cucurbits mainly depend on insects for pollination, because the male and female reproductive organs do not occur on the same flower and pollen grains are large and sticky to be carried by wind. In the present study observations on foraging behaviour of honey bee in cucumber fields indicated increase in foraging activity with the time of the day from 0600 hr to 1000 hr with maximum activity at 1000 hr which ceased after 1800 hr. Foraging activity was maximum in the second week after flowering. Foraging speed varied at different hours of the day. Bees spent more time on pistillate than staminate flowers, with maximum time spent at 1400 hr. Significantly higher fruit set, fruit weight, fruit volume, fruit length and number of filled seeds resulted when bee visits were lengthened. Similarly flower drops were inversely correlated with bee visits. It was concluded that increased bee visits to flowers results in more pollen getting transferred to stigmatic surface and also leads to better spread on stigmatic surface enabling greater entry of pollen into ovary which ultimately results in better fruit set.

INTRODUCTION

Among several cucurbitaceous crops, cucumber, *Cucumis sativus* L. has originated in India between the Bay of Bengal and Himalayas.

From India it has spread to China, Asia Minor, North Africa and Southern Europe. Now, it is extensively cultivated in tropics and sub tropics to middle part of temperate zone (Musmade & Desai, 1998). Cucumber is an important salad vegetable in Karnataka occupying an area of around 4950 hectares with total production of about 76,218 tonnes (Anon, 2001).

Fruits of cucumber vary in shape, size and colour and are considered nutritious because of their calorific value of 30, 0.4 per cent protein, 2.5 per cent carbohydrates, 1.5 mg iron and 2 mg of vitamin C per 100g fresh weight of fruit (Bose & Som, 1986).

Uniform fruit development is a primary requirement for the mechanical harvest of cucumber. Insects are required for sufficient pollination and production of uniform fruits in cucumbers. Honeybees are the primary pollinators in the field. Many times, farmers transport their bees from field to field to effect pollination. The plant being monoecious, that is, having male and female flowers in the same plant, needs pollinator assistance @ one hive of bees per 50,000 plants. The pollen of cucumbers is sticky and heavy and the flowers open only for a day. The cucumber, like other cucurbits is highly attractive to honey bees since the flowers produce large amount of nectar and/or pollen. Efficient bee management in the crop not only results in honey and wax production, but also greatly enhances the production of quality fruits and seeds through effective pollination.

Although cucumber being a highly cross pollinated, economically important vegetable crop, information regarding the role of honey bees as an input in the pollination of cucumber in increasing productivity is lacking, particularly under polyhouse conditions. The present study provides information on the foraging activity of honey bees in cucumber and the pollination success in relation to bee visits, on fruit set and some fruit characters in cucumber.

MATERIAL & METHODS

Studies on floral biology, insect pollinators and their abundance, foraging activity, quantity and quality of nectar, number of bee visits to cucumber flowers in relation to fruit set and some fruit characters, receptivity of flowers and influence of different modes of pollination on quantitative and qualitative parameters in cucumber (*Cucumis sativus L.*) was carried out at the fields belonging to U S Agri Seeds India Private Limited, Kakol, Bangalore. The cucumber variety U S 54 was grown in plot size 50 x 50 m in three plots by following all the recommended package of practices. In the present paper following observations were recorded:

Abundance of insect pollinators on cucumber flowers: The relative abundance of the different insect pollinators visiting flowers of cucumber were recorded from 0600 to 1800 hours at hourly intervals for 5 minutes during flowering period. Fortnightly observations were carried out in an area of one square meter and were expressed as mean number of pollinator visits per square meter per 5 minutes. Observations were made for three consecutive weeks following 100 per cent flowering of plants.

Speed of foraging of honey bees on cucumber flowers: The time spent by forager bee on the flower from landing till take-off was recorded by using stop watch and considered as speed of foraging. The observations were recorded from 0600 to 1800 hours at hourly intervals for 5 minutes during flowering period on both pistillate and staminate flowers. The mean time spent by each bee was expressed in seconds per flower.

Effect of frequency of honey bee visits on fruit set and fruit quality: Five pistillate flowers were selected per replication randomly and bagged using butter paper cover, a day prior to flower opening. The bagged flowers were exposed to 0, 1,2,4 and 6 bee visits and observations were made on fruit set, fruit weight, fruit volume, number of sound seeds per fruit and number of bad seeds per fruit under all the above treatments. Fruit set was expressed as percent part of total female flowers opened; fruit volume was measured by water displacement method; and sound number of seeds were the percent seeds which have ability to germinate. Bold seeds were considered as healthy. The experiment was carried out in caged condition wherein one bee colony of 5 frame strength was kept.

Foraging activity of Asis cerana indica on cucumber flowers under caged conditions: Foraging activity of *A. cerana indica* showed significant variation among different hours of the day (Table 1). The foraging activity progressed with the time of the day from 0600 hr to 1000 hr with maximum activity at 1000 hr. Activity was 11.12, 11.5 and 11.06 bees/m^2/5 min during I, II and III weeks of observations respectively. Higher activity during morning hours was due to abundant availability of pollen and nectar to the bees as the anthesis commenced at 0600 hr and by 1000 hr anther dehiscence in all the flowers would have completed, with the result plenty of pollen was available to bees at that time prompting maximum bee activity.

Table 1. Foraging activity of *A.cerana indica* (bees/m^2/ 5 min.) on cucumber flowers under caged condition

Time (hrs)	I Week	II week	III week
0600	1.12	1.06	1.06
0700	2.00	2.37	2.00
0800	3.18	3.43	3.18
0900	7.00	7.43	7.00
1000	11.12	11.50	11.06
1100	9.87	10.25	9.93
1200	8.68	9.00	8.87
1300	7.81	7.93	7.75
1400	6.73	6.69	6.12
1500	4.00	4.81	4.00
1600	2.06	4.43	3.68
1700	1.40	2.31	2.00
1800	1.43	1.81	1.43
Mean	5.31	5.62	5.24
SEM	0.24	0.25	0.28
CD@5%	0.69	0.72	0.82

Throughout the period of observation foraging activity of the bee decreased from 1000 hr onwards, and bees ceased to visit the flowers a little beyond 1800 hr. The lower activity after 1000 hr was due to depleting levels of honey.

The comparison of overall foraging activity of *A. cerana indica* on cucumber flowers during three weeks revealed that it was maximum during second week (5.6 bees/m^2/5 min) which is attributable to maximum number of flower density during this period followed by first week (5.31 bees/m^2/5 min) and third week (5.24 bees/m^2/5 min.).

There was no second peak of activity during afternoon in the caged plot unlike in the open plot. The situation within and outside the cage are quite different. The overall activity of the honey bees was more than double inside the cage than outside. Higher bee activity inside the cage following dehiscence must have depleted the forage availability faster inside than outside the cage hence the difference.

Foraging speed of A. cerana indica on staminate and pistillate flowers of cucumber: The speed showed significant variation at different hours the day. It was observed that the bee spent more time on pistillate than on staminate flowers for foraging (Table 2).

Table 2. Time spent (seconds per flower) on foraging by *A. cerana* on cucumber flowers under open conditions

Time (hrs)	Staminate flower	Pistillate flower
0700	34.55	28.50
0800	35.70	29.05
0900	37.85	29.25
1000	38.75	29.75
1100	39.75	30.87
1200	41.55	31.29
1300	42.05	30.86
1400	42.75	31.54
1500	35.75	29.50
1600	38.50	27.80
1700	36.75	27.29
1800	31.50	25.50
Mean	35.03	26.96
SEm±	00.66	00.63
CD at 5%	1.9	1.8

The maximum time was spent on foraging on staminate flowers by *A. cerana indica* at 1400 hr (31.86 sec/flower) and the minimum was recorded at 1800 hr (25.50 sec/flower). Similarly the maximum time spent for foraging on pistillate flowers was recorded at 1400 hr (42.05 sec/flower) followed by the activity at 1300 hr (42.05 sec/flower) and the minimum time spent was recorded at 1800 hr (31.50 sec/flower). However, time spent by *A.cerana indica* foraging on both staminate and pistillate flowers gradually decreased beyond 1400 hr as the day advanced.

It is obvious that pistillate flowers offer only nectar to the honey bees as opposed to staminate flowers that produce both pollen and nectar. Yield of nectar ranged from 2.05 to 2.53 µl in the former and

1.55 to 2.06 μl in the later at different hrs of the day. Nectar availability to the visitor and the variation in the nectar secretion pattern of flower may have accounted for variation in the time spent by the bee on the flower (Free, 1993)

The present findings are in conformity with the findings of Eswarappa (2001) on flowers on chow-chow and Girish (1981) on flowers of *Cucurbita pepo*, who reported that there was a difference in the time spent for collection of nectar from staminate and pistillate flowers in Bangalore. Similar results were also recorded by Mohan Rao & Suryanarayana (1988) in watermelon. The first visitor to the flower stayed longer than the later visitors and the mean time spent by it was 11.4 sec.

Effect of frequency of bee visits on fruit set and fruit characters: Number of bee visits to each flower significantly influenced the fruit set and some fruit characters such as fruit weight, fruit volume, fruit length, number of filled and unfilled seeds/fruit (Table 3). In the present study frequency of visits were regulated @ 0, 1, 2, 4, 6 per flower. Significantly higher fruit set (87.63%), fruit weight (2637.25 g), fruit volume (285.50 ml), fruit length (36.87 cm), number of filled seeds (234.75)

Table 3. Fruit set and fruit characters of cucumber in relation to frequency of honey bee visits under caged conditions.

Frequency of Honeybee visits	Flower drop (%)	Flower drop (%)	Fruit set (%)	Fruit weight (g)	Fruit volume (ml)	Fruit length (cm)	No. of good seeds
0	100.00	100.00	0.00	0.00	0.00	0.00	0.00
1	75.00	75.00	25.00	254.25	259.25	14.80	43.25
2	52.75	52.75	47.25	431.75	438.50	16.40	92.75
4	36.50	36.50	63.50	993.00	999.50	29.30	208.50
6	12.37	12.37	87.63	2637.25	1387.50	36.87	234.75
Mean	55.32	55.32	44.68	863.25	616.95	19.47	115.85
SEm±	1.89	1.89	1.54	531.86	56.72	0.78	12.12
CD at 5%	5.83	5.83	4.77	1638.96	174.8	2.41	37.35

and unfilled seeds (135.25) were recorded when the number of bee visits was increased from one to six. Whereas, with single bee visit per day per flower, fruit set (25%), fruit weight (254.25 g), fruit volume (259.25 ml) were significantly lowest while fruit length (14.80 cm), number of filled seeds (43.25) and unfilled seeds (52.75) were the least.

The per cent flower drop was significantly influenced by frequency of bee visits. With increase in bee visits there seems to be a significant decrease in flower drop. For example, all flowers dropped when bees did not visit, droppings were reduced to 75 per cent with single visit, to 36.7 per cent with 4 visits and 12.57 with 6 visits.

The present results are corroborated by several preceding findings. Eight to ten bee visits were found necessary for better fruit set in various species of cucurbits (Mc Gregor 1950; Shemotkov, 1957; Adlerz, 1960; McGregor *et al.*, 1965; Conner, 1969; Stephen, 1970). Jaycox *et al.* (1975) recorded increase in mean fruit weight as bee visits increased from one to four and twelve in cucurbits.

The significant increase in the fruit volume with increase in number of bee visits may be due to better development of fruits with higher seed set. There was a significant increase in the number of sound seeds per fruit with increase in number of bee visits which agreed with studies of Shemotkov (1957), Conner & Martin, (1969), Stephen (1970) Collison (1976), Girish (1981) and Shreeja Narayanan (2002). In their work, 8 bee visits were necessary to get good fruit set, fruit weight, fruit volume and number of sound seeds per fruit.

With the increase in bee visits it was likely that more pollen were laden on stigma, as a result, more number of ovules were fertilized which eventually resulted in better growth of fruit. This would have contributed to increased volume and weight of fruits. Adlerz (1961) reported that with the increase in bee visits to the flower there was more pollen being transferred to the stigmatic surface and also that pollen was better spread on stigmatic surface to gain entry into the ovary leading to better fruit set.

References

Adlerz, W. C.1961. Pollination of watermelon by honey bees. *Res. Rep. Flor. Agri.Exp. Stn.* 383 pp.

Anon 2001. *Statistical data on Horticultural crops in Karnataka state.* Government of Karnataka, Department of Horticulture, Lalbagh, Bangalore 13pp.

Bose, T. K. & Som, M. G. 1986. *Vegetable crops in India.* Naya Prakash, Calcutta.

Collison, C. H. 1976. *The inter-relationship of honey bee activity, foraging behaviour, climatic conditions and flowering in the pollination of pickling cucumber* (*Cucumis sativus L.*). Ph.D. Thesis submitted to Michigan State University, U.S.A. 268pp.

Conner, L. J. 1969. *Honey bee pollination requirements of hybrid cucumbers, Cucumis sativus L.* M.Sc. Thesis submitted to Michigan State University, U.S.A.150pp.

Conner, L. J. & E. C. Martin. 1969. Honey bee activity in hybrid cucumbers (*Apis mellifera L.*) *Brch. Ent. Soc. Am.* 24: 25-26.

Eswarappa, G. 2001. *Pollination potentiality of different species of honey bees in increasing the productivity of chow-chow (Sechium edule (Jacq) S.W.).* M.Sc. Thesis, University of Agricultural Sciences, Bangalore.

Girish, P. P. 1981. *Role of bees and pollination of summer squash (Cucurbita pepo Linne) with special reference to Apis cerana* F. M.Sc. Thesis, University of Agricultural Sciences, Bangalore.

Jaycox, B. R., C. Guyhin, A. M. Rhodes & J. S. Vandemark. 1975. Observations on pumpkin pollination in Illinois (U.S.A.). *Am. Bee J.* 115: 139-140.

Mc Gregor, S. E. 1950. Activity of honey bees on cantaloupe. *Rep. Iowa. St. Apiarist.* pp. 140-142.

MC Gregor, S. E., M. D. Levin & R.E. Foster. 1965. Honey bee visitors and fruit set of cantaloupes. *J. Econ. Ent.* 58: 968-970.

Mohan Rao, G. & M. C. Suryanarayana. 1988. Studies on pollination of watermelon (*Citrulus lanatus* (Thumb.) Mansf.). *Indian Bee J.* 50(1): 5-8.

Musmade, A. M. & U. T. Desai. 1998. Cucumber and melon. In: *Hand Book of Vegetable Sciences and Technology: production, composition, storage and processing* (Eds.) D.K.Salunkhe & S.S.Kadam. pp.245-272

Shemetkov, M. F. 1957. The use of bees for pollinating cucumber in hot hours and foraging bees. *Byall. Nacichnolekh Inf. Inst. Pchelovedstro.* 2: 21-24.

Shreeja Narayanan, Y. 2002. *Studies on pollination potentiality of Honey bees in Gherkin (Cucumis anguria L).* M.Sc. Thesis, University of Agricultural Sciences, Bangalore.

Stephen, W.A. 1970. Honey bees for cucumber pollination. *Am. Bee J.* 110: 132-133.

44

An Inexpensive Technique for Mass Rearing of Pod Borer, *Helicoverpa armigera* (Hubner)

SHOBHA, A., D. MANJULAKUMARI
AND GEETHA BALI

ABSTRACT

In the studies aimed at optimizing techniques for rearing a Noctuid moth, *Helicoverpa armigera*, the larva of which is a polyphagus pest on many commercial crops, influence of different larval diets on the life history of *Helicoverpa armigera* was studied under laboratory conditions. Among the three diets tested, soy bean based diet (SBBD) and chick pea based diet (CPBD) were able to support *H. armigera* larval development through adulthood. The highest quality diet in terms of larval to adult survivorship rates was CPBD (77.28%) followed by SBBD (56.18%). Corn flour based diet (CFBD) was shown to be unsuitable for development of *H. armigera* larvae as larval to adult survivorship was 0%. Using CPBD, mass rearing of *H. armigera* was carried out upto twenty six generations uninterrupted.

INTRODUCTION

Helicoverpa armigera has one of the widest distributions of any agricultural pest, occurring in portions of Africa, Asia, Australia (including Oceania), and Europe. A combination of life history traits like high fecundity, high adult mobility, polyphagy, capability for diapause, and a larval stage which is often concealed within the fruit or boll makes it a threatening pest. Its pest status is further elevated by the ease with which it develops resistance to insecticides (Srivastava, 1995) which has necessitated the need to develop alternate eco-friendly strategies for pest control. A novel approach is to use naturally occurring behavior modifying chemicals involved in the regulation of biological processes to interfere with the pest's own reproductive mechanisms resulting in reproductive failure. As a prerequisite, knowledge about the biology of the insect is necessary for manipulating its behavior. Studies on natural populations would be desirable, but, because of variability in age and densities of natural populations, these requirements are best met by rearing insects in the laboratory.

Helicoverpa species have been reared on artificial diets by a number of investigators but only few of them could be able to develop satisfactory diet formulations for continuous rearing of this moth (Sheikh *et al.*, 1989; Burton, 1969). Other reasons that may have prevented establishment of colonies are the involvement of elaborate rearing procedures because of the cannibalistic nature of the larvae and inconsistency in monitoring the rearing conditions like photoperiod, temperature, humidity and sanitation.

Since most of the research and developmental activities taking place in educational institutions have stringent budgets, the objective of the present work was to develop simple, inexpensive and satisfactory technique for establishing the moth colonies by utilizing locally available rearing materials and diet ingredients.

MATERIAL & METHODS

Parental stock: The success of the rearing program depends upon obtaining a disease-free and a genetically efficient and uniform parental stock. Therefore, *H. armigera* starter colony was established from forty pupae obtained from Biological Control Research Laboratory (BCRL), Pest Control, India Ltd., Bangalore.

Equipment required : The equipments required for rearing were B.O.D incubator, electronic balance, microwave, liquid blender, autoclave, polythene bags (autoclavable), plastic trays, rearing vials (recycled penicillin vials, standard size:5.3 x 2.3 cm^2, collected from local hospitals), oviposition containers (OC-1) (7x5x2cm^3, Bellaplast,

Germany), adult eclosion cages (dimensions:45x30x30 cm^3, mosquito-net cage), mating cages (dimensions: 11x9x9cm^3 with detachable mosquito-net), absorbent and non-absorbent cotton, forceps (blunt end), 10X hand lens, watch-glass and camel hair brush.

Conditions for moth maintenance : The eggs, adults and larvae were maintained at 25^{0}C during the photophase and at 23^{0}C during the scotophase, 65$\pm$ 5% R.H. and 12L: 12D (Light:Dark) photoperiod in a B.O.D incubator. The incubator was cleaned every week with 5% formalin.

Adult handling: The pupae obtained from BCRL were sexed and maintained in separate adult eclosion cages in a tray containing moist pre-sterilized sand bed. On emergence five males and five females were transferred to a smaller mating cage. The mating pairs were transferred into a new cage and upon separation the female was transferred into OC-1. Cotton soaked with 10% honey was placed in OC-1 as food for the adults. The female was transferred to a fresh OC-1 every day. This method though laborious, could not be avoided, as the freshly hatched larvae feed on the eggs, which would otherwise have hatched on subsequent days.

Egg handling: The OC-1 containing the eggs was checked for hatching every day and the larvae were transferred to vials containing semi-synthetic diet. The OC-1 and the mosquito net were recycled by washing in tap water and sterilizing by submerging in 10% formaldehyde solution overnight and stored until used.

Larval handling: The larvae were supported by a semi-synthetic diet. Since the larvae are polyphagus the best diet that would support the insect development and reproduction needs to be determined. The purpose of this work was to identify optimal diets for larval rearing from readily available cereal items. The diet was compounded by modifying Shorey & Hale diet (1965). The diet was based on chick pea (*Cicer arietinun*), soybean (*Glycine max*) or corn (*Zea mays*) flour.

The contents of batch I (chickpea flour/corn flour/soybean flour) (Table 1) were transferred to a domestic blender and ground thoroughly. Simultaneously agar- agar was melted in DW (batch II). The molten liquid was transferred to the blender, and the contents were mixed for 2 mins. The ingredients in batch III (Bavistint + formaldehyde 40%) were added one after another and blending was continued for 2 min. by which time, the diet temperature was reduced to less than 70^{0}C. Finally the ingredients of batch IV (Absolute alcohol and Methyl p-hydroxy benzoate) and batch V (Brewer's yeast and Vitamin syrup) were also added to the blender carefully and the contents fully whirl mixed. The final volume of the diet was made up to one litre with sterilized water.

Table 1: Contents required for preparation of approximately one litre of semi-synthetic larval diet

Batch	Ingredients	Diet Quantity
I	Chickpea flour /corn flour/Soybean flour	90.00 g
	Milk Powder	10.00g
	Sorbic acid	1.00 g
	Ascorbic acid	3.00 g
	Sterile water	400.00 ml
II	Agar-agar	12.00 g
	Sterile water	400.00 ml
III	Bavistin	2.00g
	Formaldehyde 40%	1.00 ml
IV	Absolute alcohol	10 ml
	Methyl p- hydroxy benzoate	2.00 g
V	Yeast (Brewer's)	30.00g
	Vitamin syrup	2.00 ml

Rearing vials were filled with 4ml of diet while the diet was still semi-solid and after solidification the vials were stored in refrigerator in air tight polythene bags if not used immediately.

To each vial with diet one newly hatched 1st instar larva was transferred using a camel hair brush and the vial's mouth was closed with sterile cotton wool plug. When the larvae reached the pre-pupal stage, the vials were inverted till they pupated to prevent burying of the pre-pupa in the diet.

Pupal handling: The pupae were collected from rearing vials using blunt forceps, taking care not to cause injury. The males and females were separated using a10X hand lens, and maintained separately until emergence and handled as described previously and the production cycle was repeated.

Parameters for monitoring/comparing the effect of diet on growth, development and fecundity of H. armigera: For each larval and pupal stage, the effects of a given diet on stage duration and percent mortality were examined. For adults, longevity and fecundity were determined. The data generated was subjected to Student's t-test for analysis.

Quantitative analysis of total lipids, carbohydrates and proteins in various flours used for diet preparation : The total protein was estimated by Lowry's method (Lowry *et al.*, 1951) using BSA as standard, the total carbohydrates by Anthrone method (Hedge & Hoefreiter, 1962) using glucose as standard and total lipids by van Handel (1985) method using soybean oil as standard. The data was generated by repeating the experiment thrice and was analyzed by Student t-test.

RESULTS & DISCUSSION

Influence on larval stages: Among the three diets tested, soy bean based diet (SBBD) and chick pea based diet (CPBD) were able to support *H. armigera* larval development through adulthood (Table 2). The highest quality diet in terms of larval to adult survivorship rates (77.28%) was CPBD followed by SBBD (56.18%). Corn flour based diet (CFBD) was shown to be unsuitable for development of *H.armigera* larvae as larval to adult survivorship was 0% (Table 2).

Table 2. Effect of various larval diet formulations on the juvenile developmental period (in days) in *H. armigera*

Life Stages N=600	**Duration of development with different diet formulations (in days)**		
	CPBD	**SBBD**	**CFBD**
First instar	3.35 ± 0.66	5.24 ±0.96*	2.2 ± 0.1
Second instar	2.16 ± 1.71	4.26 ±1.15*	NA
Third instar	3.13 ± 1.31	4.33 ±1.00*	NA
Fourth instar	2.65 ± 1.08	4.15 ±1.51*	NA
Fifth instar	2.48 ± 1.21	4.28 ±1.21*	NA
Sixth instar	3.45 ± 2.28	5.35 ±2.12*	NA
Pre-pupal period	2.42 ± 1.48	3.12 ±1.88	NA
Total larval period	19.64 ± 1.39	30.73 ± 1.40*	NA
Pupal period	9.94 ± 1.35	12.24 ±2.15	NA
Egg period	3.52± 0.62	3.4 1±1.15	NA

Significant difference in mean values at $p<0.05$ when CPBD was compared with SBBD, NA: not applicable

For each larval instar, the length of the instar was significantly shorter when larvae were reared on CPBD than when they were reared on SBBD (Table 1). Larval mortality during the first (22.5%), second (11.4%) and third (5.3%) instar was more in larvae reared on SBBD when compared to larvae reared on CPBD. From fourth instar onwards, percent mortality in larvae reared on SBBD was less than that reared on CPBD. But overall the mortality was significantly high in SBBD fed larvae. Irrespective of the diet, the maximum mortality was observed in the first instar (Table 3).

Table 3: Effect of various diets on the larval and pupal mortality of *H. armigera*

Life stages		**Mortality rate (expressed as percentage)**		
N=600		**CPBD**	**SBBD**	**CFBD**
Larval instars	First	9.6	22.5*	100*
	Second	1.1	11.4*	NA
	Third	1.9	5.3*	NA
	Fourth	2.6	2.3*	NA
	Fifth	2.32	1.1	NA
	Sixth	1.54	0*	NA
Pupae	3.66	1.22	NA	
Total mortality	22.72	43.82*	100*	

- Significant difference in mean values at $p<0.05$ when CPBD was compared with SBBD, NA: not applicable

The time taken for adult emergence from pupae reared on SBBD was 2.3 days more than those fed with CPBD but this difference was insignificant (Table 2).

Influence on adult life: It was observed that all the moths were able to reproduce on CPBD and SBBD diets. Preoviposition period of both mated and virgin females was one day more in adults that eclosed from larvae fed on SBBD when compared to adults that eclosed from larvae fed with CPBD (Table 4). There was no significant difference in time of egg hatching between these diets (Table 2).

Irrespective of the diet provided, the number of eggs laid by mated female was significantly more than the virgin female. Estimates of mean total lifetime fecundity as measured by the number of eggs laid plus number of matured eggs retained in the abdomen differed significantly between the two diets. Lifetime fecundity was highest when females were reared and maintained on soybean diet. The adult longevity was not affected significantly by any of these diets except among virgin females which lived longer on SBBD. Comparatively, mated male and female adults lived for a shorter period than the virgins (Table 4).

Table 4: Effect of various diets on preoviposition period, fecundity and adult longevity of *H. armigera*

	N=50 for each group	CPBD	SBBD
Preoviposition	Mated female	1	2
period in days	Virgin female	1	2
Fecundity	Mated female	1986± 64	2265 ± 58*
(Mean±SD)	Virgin female	488 ± 26	528 ± 36 *
Adult longevity	Mated female	8.5 ± 1.5	9.1 ± 0.2
in days	Virgin female	14.5 ± 0.5	16.5 ± 0.1 *
(Mean±SD)	Mated male	13.25± 0.5	14.75± 0.9
	Virgin male	14.2 ± 0.2	16.3 ± 0.7

Significant difference in mean values at $p<0.05$ when CPBD was compared with SBBD, NA: not applicable

Nutrient levels in the various flours used for larval diet preparation: The total protein content was highest in soybean and least in corn flour, while chickpea contained moderate amount of proteins. Total carbohydrates and total lipid concentrations were moderate in soybean, but highest in chick pea and least in corn flour. It is clear from the results that corn flour had the least concentration of all the estimated nutrients (Table 5).

CPBD and SBBD were better in nutrient quality compared to CFBD but mortality rate was less in CPBD. Hence CPBD was used as larval diet for rearing *H. armigera* in our laboratory. The larvae were reared successfully through 26 generations from March 2003 to April 2006 on chick pea based diet from parental stock of forty pupae.

Larval food plays an important role in the establishment of insect population in laboratory. It not only determines build-up of population by influencing growth and development during larval stage but also greatly influences longevity as well as fecundity during adult stage (Lukefahr & Martin, 1964). The larval diets used in these investigations were based on the diet developed by Shorey & Hale (1965). The modifications made were incorporation of casein, commercially available multivitamin syrup and the use of either soybean/chickpea/ corn flour in place of pintobean. The other modifications made were use of locally available recycled materials like penicillin vials for larval rearing, disposable food containers for oviposition. For adult diet, honey (BCRL, PCI., India) was used. Previous studies have shown that incorporation of sorbic acid inhibited growth of yeast and molds (Mardaus & Buckner, 1997). All the diets formulated were devoid of fungal development for 30 days, but if the faeces of the larvae mixed with the diet, fungal growth was observed after 20 days. The most commonly observed fungus was *Aspergillus niger*.

Table. 5. Content of protein, lipid and carbohydrate in flours used for larval diet preparation

	Concentration (mg/g)		
	Chickpea flour	**Soybean flour**	**Corn flour**
Protein	340.45±31.21[a]	482.35±21.43[b]	102.12±23.67[c]
Carbohydrate	418.46+35.14[a]	213.32+26.12[b]	52.13+12.42[c]
Lipid	31.14+09.12[a]	17.68+02.10[b]	06.24+13.10[c]

(Mean values followed by different letters in the same row are significantly different at $p<0.05$)

The present study showed that the larvae of *H. armigera* survived on chickpea and soybean based diets, though survival rates greatly varied, reflecting their polyphagus nature, but none of the larvae survived on CFBD. CPBD appeared to be the best diet as the larval to adult survivorship rates was highest among the diets tested. The larval mortality was highest in the first instar for all the diets similar to observations made by Kyi & Zalucki (1991) but subsequently in the later instars, the mortality decreased. Although the larval size was not measured, we observed that the larvae which died in the late larval stage were those which had smaller body size and longer larval period. Studies in the past have shown that the complete larval period lasted between 12-36 days on different diets (Kirkpatrick, 1962; Bhatt & Patel 2001; Fowler & Lakin, 2001). It was found that the larvae of *H. armigera* fed on either corn silks or tender corn cobs had mean larval periods of 21.8 and 22.2 days (Coaker, 1959), and it was 18-21 days for those fed with different varieties of chickpea (Sanehdeep Kaur *et al.*, 2002) while for those fed with immature soybean seed, it was 23 days (Jallow *et al.*, 2001). The results in the present study are very close to the field study with respect to the larval period as far as CPBD (19.65 $\pm$ 1.39 days) is concerned. While on soybean diet larval development was slow (30.73 $\pm$ 1.40 days) when compared to CPBD, none of the larvae survived on CFBD. This is probably because of the low levels of nutrients like protein, carbohydrates and lipids in the corn flour compared to the other flours indicating that incorporation of corn flour in artificial diet is not feasible for rearing *H. armigera* larvae despite the fact that they grow very well on maize cobs in the field.

In nature, larvae move between 2.5-17.5 cm below the soil surface to pupate depending on soil moisture, organic matter on the surface, and other factors (King, 1994). Less frequently, pupation occurs within a spun web on the host plant (e.g. in a corn cob) or on the soil surface. Depending on temperature and nature of larval diet, the pupal stage lasts between 6-33 days, unless the insect goes into diapause, in which case pupation may require several months (Fowler & Lakin, 2001; CAB 2003). We observed that the pre-pupae burrowed into the diet and sometimes died due to asphyxiation if the vials were not inverted. The pupal period was found to be less in larvae fed with CPBD when compared to the larvae fed with SBBD.

In our study, the total carbohydrate and lipid was found to be more in CPBD which could probably have influenced juvenile development. Diapause was not observed in pupae developing from larvae fed on any of the diets, probably because the insects were maintained at 25 ^{0}C and 23^{0}C during the photo- and scotophase respectively. Several adult longevity studies report a range in adult life span of 5 to 36 days depending on mating status, larval and adult diet and temperature (King, 1994). The mated female lives for very short duration of 8.5 $\pm$ 1.5 and 9.1 $\pm$ 0.2 days when reared in CPBD and SBBD respectively, but, both mated and virgin males lived for a similar period. SBBD influenced both virgin male and female moth's longevity indicating that the content of protein in the larval diet greatly influences longevity of the moths.

A protein diet or protein stored in the fat body is required for egg production in social wasp, *Polistes canadensis* (Giray *et al.*, 2005). In

H. armigera also we found that soybean which is a rich source of protein greatly influenced fecundity.

Present results clearly suggest that the quality of the diet greatly influences the growth, development and fecundity of *H. armigera*. While increased carbohydrate and lipid influences the preimaginal development, incorporation of protein is necessary for maximum fecundity. Hence it was concluded that the CPBD based diet tested during the course of these studies was found to be most suitable for mass rearing of *H. armigera*.

Presently we are developing an artificial diet by combining both chickpea and soybean. It is also interesting to observe that we can delay or accelerate the development of this insect by changing the diet ingredients qualitatively which would greatly help in obtaining insects of different larval instars on the same day from the same parental stock, thus preventing intraspecific variation.

ACKNOWLEDGEMENTS

We would like to thank the Department of Science and Technology, Government of India, for providing financial assistance.

References

Bhatt, N. & R. Patel. 2001. Biology of chickpea pod borer, *Helicoverpa armigera. Indian J. Entom.* 63: 255-259.

Burton, R. L. 1969. Mass rearing the corn earworm in the laboratory. *USDA-Agricultural Research Services Series.* 133-134.

Cab. 2003. *Crop protection compendium: global module.* Commonwealth Agricultural Bureau International, Wallingford, UK.

Coaker, T.H. 1959. Investigations on *Heliothis armigera* Hb in Uganda. *Bull. Entom. Res.* 50:487-506.

Fowler, G. & K. Lakin. 2001. Risk Assessment: The Old Bollworm, *Helicoverpa armigera* (Hubner), (Lepidoptera: Noctuidae). USDA-APHIS, Center for Plant Health Science and Technology (Internal Report), Raleigh, NC. pp. 1-19.

Giray, T., M. Giovanetti & M. J. West-Eberhard. 2005. Juvenile hormone, reproduction, and worker behavior in the neotropical social wasp, *Polistes canadensis. Proc. National Association for Science.* 102:3330-3335.

Hedge, J. E. & B. T. Hofreiter. 1962. In: *Carbohydrate chemistry 17* (Eds.) Whistler, R. L. & J.N.BE. Miller. Academic press, New York.

Jallow, M. F. A., M. Matsumura & Y. Suzuki. 2001. Oviposition preference and reproductive performance of Japanese *Helicoverpa armigera* (Hübner) (Lepidoptera: Noctuidae) *Appl. Entom. and Zool.* 36: 419-426

Kirkpatrick, T. H. 1962. Notes on the life-histories of species of *Heliothis* (Lepidoptera: Noctuidae) from Queensland. *Queensland J. of Agr. Anim.Sci.* 19: 567-570.

Kyi, A . & M. Zalucki. 1991. An experimental study of early stage survival of *Helicoverpa armigera* (Lepidoptera: Noctuidae) on cotton. *Bull. Entomol. Res.* 81: 263-271.

Lukefahr, M. J. & D. F. Martin. 1964. The effects of various larval and adult diets on the fecundity and longevity of the bollworm, tobacco budworm and cotton leaf worm. *J.Econ.Entomol.* 57: 233–235.

Lowry, O. H., N. J. Rosebrogh, A. L. Farr & R. J. Randall. 1951. Protein measurement with the folin phenol reagent. *J. Biol. Chem.* 193: 265-275.

Mardaus, M. C. & J. S. Buckner. 1997. Identification and characterization of novel short-chain acid esters as surface lipid components of *Heliothis virescens* pupae: Effect of sorbic acid in the diet. *Insect Biochem. Molec. Biol.* 27:551-561.

Sanehdeep Kaur, K.S. Chhabra & B. S. Arora. 2002. Larval development of *Helicoverpa armigera* Hubner on wild species and cultivated genotypes of Chickpea. *J. Insect Sci.*15:67-69.

Sheikh, R. M., D. Sheikh, S. Baqir & S. Naqvi. 1989. Mass rearing of corn Earworm *Heliothis armigera* (Hubn) on modified Noctuid diet. *J. Zool.* 21:115-121.

Srivastava, C.P. 1995. Insecticide Resistance in *Helicoverpa armigera* in India, In : *Resistant pest management* Web edition ; Vol. 7, No. 1.

Van Handel, E. 1985. Rapid determination of total lipids in mosquitoes. *J. American Mosquito Control Assoc.* 1:302-304.

45

Diverse Feeding Trends in A Primitive Oribatid Mite, *Archegozetes longisetosus* Aoki (Acari: Oribatei)

SHIJI, M.T. AND N. RAMANI

ABSTRACT

Archegozetes longisetosus, a ubiquitous parthenogenetic oribatid mite species with highly diverse nutritional habits has been selected to study its unusual feeding trends, if any, and also to assess its exact role in biodegradation. The large body size of the mite coupled with its short life span and diverse feeding habits make it ideal for experimental studies. Results of feeding studies often supplemented highly inconsistent results on the behavioural pattern of the mite.

Live mites for the study were recovered through Berlese Tullgren funnel extraction of the litter and soil samples collected from rubber plantations at Mullankunnu, Calicut district of Kerala, India. The live mites were extracted into vials containing moistened leaf powder. Culturing was carried out in plastic vials on a substrate comprised of a mixture of plaster-of-paris and charcoal (4:1 ratio). For feeding experiments, an array of food materials comprising 7 species of fungi, yeast, mushroom, leaf litter, boiled rice, cotton thread, artificial sponge, filterpaper, moss and live animals like earthworm and millipede larva etc. were offered to know the feeding response of the species to individual food items. Results of the feeding studies revealed the ability of the species to consume all the test food items. However fungi like *Curvularia geniculata, Pestalatia* spp, *Alternaria alternata, Colletotrichus* spp. and yeast and leaf litter were the most favoured food items. An unusual feeding trend was the zoophagous habit of the species, in which the adults and nymphs showed voracious feeding on live earthworm and millipede larva. Another fascinating observation made during the study was the cannibalistic behaviour exhibited by the adults, larvae and immatures of the species. The adults voraciously consumed the larva and all the three nymphal stages while the larvae and nymphal stages exhibited voracious feeding among themselves, excluding the adults. Thus the species apart from its saprophagous feeding mode, possesses predatory and cannibalistic habit also which are so far not reported among members of the genus *Archegozetes* and the family Trihypochthoniidae.

SECTION - 5
APPLIED BEHAVIOUR

46

Prevention of Bird Damage to Maize Cobs

SHIVASHANKAR, T

ABSTRACT

Maize (*Zea mays*) is an important crop grown in semi-arid and rainfed areas of Mandya and Mysore districts of Karnataka. The crop is grown in isolated patches and attracts a large number of Roseringed Parakeets. Even by following recommended practices the crop is not protected from bird damage. Farmers are incurring crop losses ranging 14 to 40 %.

Field trials were conducted in and around Nagenahalli, Mysore to protect maize cobs from bird damage. New methods were developed to protect maize cobs from bird damage. Among granivorous birds, Roseringed Parakeet (*Psittacula krameri*) is the major species damaging crop. The trials on cob protection from birds were conducted in farmer's fields. Cobs were covered with nylon bags/shining paper covers of different colours. There was difference in the damage between cobs covered and uncovered. The method was very effective for longer periods. Even in larger areas, cobs covered with nylon net and shiny paper bags received no or very little damage. This mechanical method can form an important tool in the management of bird damage. Timely harvest of cobs, watch and ward and scaring can supplement the effect of covering cobs.

INTRODUCTION

Maize (*Zea mays*) is one of the important cereal crops in Southern Karnataka, India. Many farmers grow this crop in small isolated patches, sometimes larger areas also. This crop attracts many avian granivores especially Roseringed Parakeets (*Psittacula krameri*) which cause heavy losses resulting in 3 to 20 % yield loss in maize (Ramzan & Toor, 1972; Dhindsa *et al.*, 1992; Rao & Dubey, 2006). Wrapping maize cobs with maize leaves, reflective ribbons, bioacoustics, pyrotechniques, screen crops, spraying botanicals and habitat management are some of the management tools available to reduce bird depredation on crops. However, these tools are not popular/ properly adapted by the growers. Many of them are time consuming, effective for short period or involve high cost and labour intensive. Some of the tools are not effective in long run as the birds get habituated to them gradually (Rao & Dubey, 2006). Therefore, there is a need to develop alternative/conventional methods that are effective, economical and ecofriendly. Hence, the present study was conducted at Mandya and Mysore districts of Karnataka.

METERIAL & METHODS

Experiments 1 to 3 were conducted at Agricultural Research Station, Nagenahalli, Mysore, Karnataka. All the experiments were carried out from the milky stage (65 days after sowing) to harvesting stage (120 days) of maize variety NAC 6004. The experiments were restricted to the peripheral (border) rows (Rao & Dubey, 2006). Experiment 1 was conducted during September–October, 2003. Red and yellow (n= 50) shining paper measuring 10x15 cm were used in all the experiments of the study. Two centimeter deep cuts were made at about every 5 mm of the paper. The colored paper strips were placed on each row of the cobs so that the cut end faced upwards. Two rubber bands were used on each cob to secure the paper.

In another plot white and orange (n=50) colored nylon nets of 25 cm long were used. The nylon nets were placed from the top of the cob after removing the dried silk. One rubber band on the middle of each cob was placed as control. Observations on the number of cobs damaged by Roseringed Parakeets in the treated and control plants were recorded at regular periods.

Experiment 2 was conducted during 2004-05. The north and eastern borders of the experimental plot had roads with tall trees. Observations on the number of cobs damaged by birds were recorded during 2004 on both the borders. During 2005, along the eastern border 166 cobs out of 756 were covered with red, pink, blue and yellow shining papers. In the control plot, rubber bands were placed on 175 out of 832 cobs on the northern border.These treatments were carried out prior to the commencement of the damage by birds. Observations on the number of cobs damaged by the Roseringed Parakeet in the treated and control plots were recorded at regular intervals.

Experiment 3 was conducted during September–October 2005. In this plot the maize cobs present along the two peripheries (north and eastern borders) were observed every day in the morning for fresh feeding symptoms by the Roseringed Parakeet. The cobs that showed initial feeding by the birds were immediately covered with coloured shining paper (n=794) on the eastern border. Red, blue, yellow, silver and pink shining papers were used for this experiment. Choice of placement of different colors was random. In the control plot, rubber bands were (739) used on the northern border. The number of cobs damaged by the Roseringed Parakeet in the treated and control plots were recorded at regular intervals.

Experiment 4 was conducted during August–September 2004 in a farmer's field at B.K. Doddi, Malavalli taluk, Mandya district, Karnataka. Red, pink, blue and yellow shining papers (each color on 400 cobs) were used for this experiment. These shining papers were placed along the three borders of the crop. Rubber bands were placed as control on 400 cobs along the fourth border. Observations on the number of cobs damaged by the Roseringed Parakeet in the treated and control plants were recorded at regular intervals.

RESULTS & DISCUSSION

No damage/feeding by Roseringed Parakeet was observed on the cobs covered with shining papers from 7th to 21st day of paper coverage. Majority of the nylon nets covered cobs were also free (95%) from the bird damage. Both the shining papers and the nylon nets were significantly superior ($P<1$) to the control plants in protecting maize from bird damage (Table 1). Amongst the white and orange colored nylon nets, the latter was superior. In few nylon netted cobs, the

Roseringed Parakeets had tried to open/cut the husk (outer cover of the cobs) but were unable to feed on the grains. Bird damage was comparable (24 and 26 %) on both the sides during 2004 (Table 2). The border row covered with colored shiny papers (25% of the cobs were covered) was significantly less damaged by birds compared to the control (border 2) as well as to the previous year's damage (Table 3).

Results of the experiment 3 clearly indicated that there was significant reduction in the cob damage when cobs were covered with shining papers at border row 4 (3%) when compared with uncovered cobs (25%) at border row 2 (Table 3). In experiment 4, where four different colored shining caps were wrapped, there was no difference in the damage; all the four methods prevented bird damage equally.

Table 1. Effect of colored shining papers and nylon nets in deterring the feeding of Roseringed Parakeet on maize cobs.

Treatments	Number of cobs	7th day of treatment	14th day of treatment	21st day of treatm-ent	% Damage
Nylon netted white bags	50	0	2*	4*	8
Nylon netted orange bags	50	0	0	1*	2
Red shiny paper	50	0	0	0	0
Yellow shiny paper	50	0	0	0	0
Control	50	5	16	21	42

*Damage attempt by birds was seen on these cobs but were unable to feed on the seeds

Table 2. Effect of different colored papers placed at border plants before commencement of bird damage.

Treatment	Total plants in the border row	No. covered	No. damaged	% Damage
Border 1 (year 2005)	756	166	4	0.5
Previous year(2004)	882	-	239	26
Border 2 (year 2005)	802	0	287	36
Previous year (2004)	832	-	198	24

Table 3: Effect of placing shining papers after the birds started feeding on the cobs

Treatment	Total plants	No.covered	No. damaged	% Damage
Border 1	794	234	21*	3
Border 2	739	0	187	25

* damage was observed only in the uncovered cobs along the border.

Table 4. Effect of caps of shining paper to prevent bird damage at B.K Doddi, Malavalli, Mandya (Sept-Oct. 2004).

Treatments	Number of cobs*	10th day of treatment	26th day of treatment
Red shiny caps	400	0	0
Pink shiny caps	400	0	0
Blue shiny caps	400	0	0
Yellow shiny caps	400	0	0
Control	400	8+	110+

*= None of the cobs were damaged due to birds. + No. of cobs damaged
63 caps had fallen by 26th day due to breakage of rubber bands.

Roseringed Parakeets are the major bird species damaging maize cobs particularly during milky stage. Traditionally maize cobs were prevented from bird damage by wrapping with leaves of maize in addition to scaring and watch and ward. Though these methods are cheaper and eco friendly, birds gradually get habituated and try to puncture the leaves and cause damage. Present day cultivation of maize is facing lot of labour problem and also involves high investment. Different colored shining papers were more effective when compared with nylon netted color bags. These shining caps prevented bird damage by repelling birds due to their reflective nature. Rao & Dubey (2006) have shown that indigenous methods are sometimes less effective due to the habituation of birds. Reflectors (ribbons etc) proved very effective against birds damage to maize cobs as there is no habituation of birds and less cost effective also (Ramzan & Toor, 1973). Dhindsa *et al.*, (1992) conducted similar trials and showed that wrapping caps with shining paper are very effective over long periods and less cost effective.

CONCLUSIONS

Among avian granivorous birds, Roseringed Parakeet (*Psittacula kramari*) is the major bird pest, damaging maize cobs and sometimes causing economic loss if there is no timely protection. The per cent damage may vary from place to place depending on the surrounding habitat. Findings of the present study, i.e., covering the maize cobs with colored shining paper caps proved very effective for long period, economical and eco friendly too. Hence, the farmers can very well adopt this method to protect maize crop to a grater extent from bird damage.

References

Dhindsa, M. S., H. K. S. Saini & H. S. Toor. 1992. Wrapping leaves around cobs to protect ripening maize from roseringed parakeets. *Trop.Pest.manage.* 38: 98-102.

Ramzan, M. & H. S. Toor. 1973. Damage to maize by roseringed parakeet *Psitaacula krameri* in Punjab. *J.Bombay Nat. Hist.Soc.* 70: 201-204.

Rao,V. V & O. P. Dubey. 2006. Granivorous bird pests and their Management. *In: Vertebrate Pests in Agriculture – The Indian Scenario (Ed.)* Shakunthala Sridhara. Scientific Publishers, Jodhpur. pp.29-61.

47

Biodiversity-A Biologist's Perspective

KRUPANIDHI, S.

ABSTRACT

The myriad forms of living beings along with their habitats constitute the Biodiversity. The molecular diversity is manifested in the form of species diversity. The same along with their respective habitats is rephrased as biological diversity. The present day realization is that the components of biological diversity are in coexistence and contributing for the wealth of the ecosystem and the same has led to the widely prevalent usage of the term "biodiversity". Of late, human intervention into the natural ecosystems is daunting and bringing alterations in the balance between the different biotic components. Habitat preservation is one of the best ways to conserve biodiversity. The conservation of biodiversity and restoring ecosystems require a balanced approach to manoeuvre the situation, which calls for the study of systematics, ecology, behaviour and informatics to appreciate the adaptability of organisms in their respective habitats. Eventually, a biologist perceives biodiversity as an outcome of an evolutionary stable strategy nurtured by Nature.

INTRODUCTION

The term 'biodiversity' is now being used as a buzz-word. It is a recent phrase coined by W. G. Rosen in 1985. Everyone, viz., legislators, administrators, industry, nongovernmental organizations and general public are showing interest on varied aspects of conservation of biodiversity and in culturing and cultivation of elite species. Much is being spoken on the steps to be taken on the conservation of biodiversity (Gasten & Spicer, 2004) ignoring the fact that there is an unending ignorance among the people about the diversified life-bearing forms around them. Hence, it is an issue of paramount importance to recognize the importance of experts in biology to impart the principles of biological sciences for understanding the significance of diversified flora and fauna. An account of the following provides ample scope to develop multidimensional conservation policies: (1) Components of biodiversity (2) Avenues to brood over (3) Role of systematics in conservation issues (4) Coexistence (5) Biodiversity and informatics (6) Evolution of molecular diversity and (7) Suggested strategies.

(1) Components of Biodiversity

Sex produces diversity. Diversity is the propeller of evolution. The flora and fauna, with dazzling variety of colours, shapes and behaviour around us belong to varied species inhabiting natural habitats. They are the products of biological evolution. That is to say that these organisms withstood the test of Natural Selection and flourished. The gene is the fundamental unit of Natural Selection. The genetic diversity is the diversity of genes within a species (Krupanidhi, 2005). Consequently, there is genetic variability among the members of a population and the individuals of the same species. The genetic variability implies that the vulnerable individuals of the population are susceptible during competition. The species diversity is the diversity among species in an ecosystem. Biodiversity hot spots are excellent examples of species diversity. Ecosystem diversity is the diversity among the components of an ecosystem. The present day conservation biology is attempting to conserve the natural ecosystems. Biodiversity is usually evaluated as taxonomic richness of a geopgraphic area encompassing species richness or evenness or extinction. There are three indices which are used by ecologists: Alpha diversity is witnessed within an ecosystem and is measured by counting the number of species. Beta diversity is species diversity between ecosystems and this involves comparing the number and types of taxa that are unique to each of the ecosystems. Gamma diversity is a measure of the overall diversity of different ecosystems within a region (Biodiversity-Wikipedia). The behavioural aspects of Biodiversity is the diversity of durable cooperative interactions among species and also in between species and biotope and species and biome.

(2) Avenues to brood over

- Why contemplate on Biodiversity, Ecology and Conservation Biology?
- What is the greatness of Biodiversity?
- Do we need to protect Biodiversity?
- What is the role of biologists in conserving Biodiversity?
- Is there a need to disseminate the knowledge of Biodiversity?
- Why is Biodiversity important?
- Are we crunching the natural habitats due to need or greed?
- What makes our environment?

(3) Role of systematics in conservation issues

The richness of species diversity, taxonomic importance of transient species, the vulnerability of a few species due to non-adaptability, commercial exploitation of elite species, the negative effects of inbreeding and consequent genetic homogenization-the knowledge of all these is of immense importance to multi-dimensional conservation policies. The biological systematics deals with the study of the diversity of life on earth and the interrelationships among the members of biota (Mayr & Ashlock, 1991). It facilitates the understanding of evolutionary history of life on earth. Today, the science of systematics utilizes the tools and techniques of molecular biology, recombinant DNA technology (RFLP, AFLP, SNP, etc) and computer programmes (PHYLIP, PAUP, MEGA-3, Clustal W etc). There are about 100 million species on the earth. The evolution, speciation, distribution, dominance and susceptibility of these species provide an important baseline for studies of conservation and recording of endemic diseases. Taxonomy is the theory and practice of classifying organisms. Carolus Linnaeus (1707-1778) is the father of modern taxonomy. He grouped species according to shared physical and morphological characteristics. These groupings have been revised to support and improve the Darwinian principle of common descent. The classification is the method by which biologist categorises organisms into taxa by deductive and inductive procedures. Currently, molecular systematics and digital taxonomy are aiding in making diversity accessible at nooks and corners of the world. A biological species is a 'default' reproductive unit perfected by Natural selection and continuously subjected for biological evolution in a low pace resulting into a transient species. Conservation steps undertaken for protecting endangered species include cryopreservation of gametes, assisted reproduction, captive breeding and somatic cell hybridization to propagate transient and endangered species both *in situ* (in nature) and *ex situ* (intensively managed programmes in captivity).

Numerical phenetics: The phylogenetic trees, the principles of animal taxonomy and tools of classification provide us the insights into the grouping and origin of taxa. However, these tools do not conceive the method for highlighting interrelationships among them. The numerical phenetics is one of the tools to evaluate the distance of relationships among the groups of species in a taxa. For example ten families of the Order Primates can be short-listed by virtue of the presence of characters that show at least three notable variations amongst the taxa. These characterstics are numbered and analyzed using the Manhattan Distance method of Numerical Phenetics. From the similarity coefficient units obtained thereby, the families can be drawn closer and clustered into a dendrogram. The chosen families of the Order Primates come within the ambit of a dendrogram comprising of several primary, secondary and tertiary clusters. These clusters demonstrate the distance of relatedness. A single family viz., *Soricidae*, of the Order Insectivora can also be included to determine its relation with respect to the primates. The position of it in the dendrogram indicates the common ancestry for all the primates from the insectivore stock, based on their similarity coefficients (Viswanathan & Krupanidhi, 2006).

(4) Coexistence

The Egyptian hieroglyphic depicting the sun, the serpent and the wings reflect that life is a biological phenomenon and essentially involves the master plan of Nature and the same is well depicted here, i.e.: The light emanates from the Sun and travels into the universe as energy. The energy is symbolized as serpent. Energy, however, cannot exist on it's own. It needs a substratum and hence percolates through the medium of life and is constantly connected back to its source by wings. Soaring wings symbolise love. The flow of energy through all the living beings is an indication that life exists through the phenomenon of interdependence.

In the present genomic era, there is an incessant and enhancing scope in exploring biodiversity. Therefore, there is an upsurge in understanding the cooperative existence of living beings of diversified groups and the causative agents for the same, such as the habitat, sharing of genes, behaviour, receptivity, competition, territorality etc.

Charles Darwin (1859) and his contemporary field naturalist, Alfred Russel Wallace were influenced by reading Malthus's "Essay on Population". Darwin was struck by an idea in Malthus's writings. Malthus had said that population multiplies faster than food. If that is true of animals, then they must compete to survive, so that nature acts as a selective force, killing the weak, and forming new species, which are well adapted to their environment. These contemporary thoughts made Darwin to use very strong phrases in his concept of Natural Selection viz., struggle for existence, survival of fittest and prodigality of production (Barnett, 1958). Darwin did notice that the bearers of a trait must themselves benefit if the trait was to be favoured under Natural Selection. He erroneously recognized that Natural Selection could not favour a trait in one species solely for the benefit of another species. Natural Selection cannot possibly produce any modification in any one species for the good of another species (Darwin, 1859). Thus, the evolution of cooperation was not understood during Darwin's time. However, the emphasis was focused on the adaptability and evolution of traits. Furthermore, the adaptability of an organism is not only with the environment but also with the neighbours sharing the same habitat (Sachs Joel *et al.*, 2004). However, the biological interactions and molecular interplay among the species bestowed the mechanism of cooperation since time immemorial and the same is the crux for the sustenance of biodiversity.

(5) Biodiversity and informatics

Of late, systematists are making extensive use of molecular genetics and computer tools (Stanley, 2000; Krupanidhi, *et al.*, 2007) to study organisms and their habitats. Digital Taxonomy provides a range of links on software methodologies and data sources. There are 200 open source biological software websites available. Eg. Fishbase, Genbank, LegumeWeb, etc. In this section, it is intended to put forward one of our research investigations on the use of *in silico* tools in evaluating the interleukins (IL) of the hematopoietin family of cytokines. The focus is laid primarily on the consensus homology in relation to the interleukin sequence alignments and biological functions. First, we performed an interspecies IL-2 protein comparison among a few chosen members of the Class Eutheria (true placental mammals). Results precisely confirmed the homology between humans and chimpanzees as the closest related class members. In addition, *Macaca* constituted an early offshoot of the primary cluster. Thus, the distance of relationships among the groups of taxa can be evaluated. Next, the related sequences of IL-2 through IL-9 of *Homo sapiens* were obtained by NCBI and/or BLAST search. Upon submitting these related IL sequences on-line to the ClustalW software, there appeared a rooted tree, viz., cladogram, which had two primary IL clusters, and a multiple sequence alignment. Interestingly, each primary cluster pair had redundant biological functions. Furthermore, the derived phylogenetic tree suggested a hypothetical ancestral origin, an early period of gene duplication, and an extended period of divergence and specialization among all the inteleukins analysed. In the amino acid sequence alignment, residue analyses among chosen IL family members in humans revealed that there was relatively limited homology; however, there was hydrophobic residue conservation and preserved secondary structure consensus. Finally, a protein data base (PDB) search of ILs allowed a three dimensional analysis of IL-6 using Rasmol, which revealed 5 á-helices with predicted amphipathic faces. Thus, *in silico* method of analysis allows biology students to interact with online tools to enquire about the molecular and biological functions.

(6) Evolution of molecular diversity

Evolution involves the transformation and splitting of lineages. The genetic differences among species are used to reconstruct their evolutionary history or phylogenetic tree. The amino acid sequence for cytochrome C in a variety of organisms revealed that it has changed very slowly during evolution. Cytochrome C sequences between human and chimps are identical; there is a difference of one amino acid between humans and rhesus monkeys denoting their divergence from a common ancestral species approximately 20 million years ago. The number of amino acid differences between Cytochrome C in humans and a variety of other organisms are as follows: man is closely related to other mammals than are insects, and men are more closely related to insects than are to fungi. Like wise, Cytochrome C of man differs in 13 amino acids from that of dogs, in 36 amino acids from that of moths, and in 56 amino acids from that of yeast. More than one nucleotide change may be required to change a given amino acid. Thus, the *minimal mutational distance* between the genes of any two species is established by totaling the change in the nucleotide substitutions of the amino acids replaced in a protein due to mutations. The result is the evolutionary tree or phylogeny that unites the species. The extant species are connected to the inferred common ancestors, which are still connected to earlier common ancestor, culminating in a single common ancestor for all the species on the tree. The molecular diversity of respiratory pigments viz., haemoglobin, hemerythrin, and hemocyanin is another classical example to visualize their evolution and reaching perfection in their adaptation i.e., cooperative binding with oxygen (Krupanidhi, 1988). They are functionally homologous and structurally diversified. The primordial alpha helices of proteins must have independently coupled with heme group to develop into myoglobin; and the same through fusion possibly has led to a compact tetramer namely haemoglobin molecule. The Fe and Cu as the prosthetic groups must have bound independently with the primordial alpha helices of proteins and possibly underwent the fusion and duplication which must have led to the building up of large sized molecules namely, hemerythrin and hemocyanin. They function in oxygen transport with limited geospatial distribution. The immunoglobulins of vertebrates are yet another classical example of molecular diversity nurtured by Nature.

(7) Suggested strategies for conserving Biodiversity include

- Inherently the fauna and flora always protect environment. It is wise to adopt those strategies to sustain the biodiversity. Largely, animals and plants neither cause damage nor intuitively work to improve the vicinity.
- Humans are psychosomatic entities. They are endowed with societal attributes and also blessed to do fine work incessantly to protect their surroundings.
- Dissemination of the knowledge on biodiversity to the public.
- Keeping the public aware of the damage to ecosystem caused by humans.

References

Barnett, S. A.1958. *A Century of Darwin*. Harward University Press, Cambridge, Maschusetts, USA.

Biodiversity : Wikipedia, the free encyclopedia, Online search.

Darwin, Charles. 1859. *The Origin of Species*, A Mentor Book Publication: New American Library.

Gaston, K. J. & J. I. Spicer. 2004. *Biodiversity, an introduction*. 2nd Ed. Black Well Publishing.

Krupanidhi, S. 1988. Respiratory pigments. *Biol. Edu.* April-June pp104-113.

Krupanidhi, S. 2005. Designer genes. *Everyman's Science*. XXXIX 5: 304-308.

Krupanidhi, S., S. S. Madhukar & S. D. Doughman. 2007. *In silico* tools for evaluating the consensus homology among interleukins. *Online J. of Bioinformatics*. 8 (1):1-6.

Loewenberg, B. J. (Ed). 1959. *Charles Darwin: Evolution and Natural Selection*, Beacon Press, Boston.

Mayr, E. & P. D. Ashlock. 1991. *Principles of Systematic Zoology*. Second edition, McGraw Hill international Edition, New York.

Sachs Joel L., G. M. Ulrich, T. P Wilcox & J. J. Bull. 2004. The evolution of cooperation. *The Quart. Rev. of Biol.* 79 (2): 135-160.

Stanley D. B. 2000. *An overview of biodiversity informatics*. Wikipedia.

Viswanathan, A. S. & S. Krupanidhi. 2006. Numerical phenetics–A dendrogram for primates. *Proc. Zool. Soc. India* 5(1):1-9.

48

Validation of Integrated Pest Management Modules for the Management of Shootfly in *Rabi* Sorghum

BALIKAI, R. A.

ABSTRACT

A field trial with various Integrated Pest Management components was conducted at the Regional Agricultural Research Station, Bijapur to find out low cost means for the management of shootfly in *rabi* sorghum. Early planting with normal seed rate (8 kg/ha) proved to be significantly superior over late planting with high seed rate (10 kg/ha) with respect to reduction in shoot fly incidence and harnessing higher seed yield irrespective of main treatments. The main treatment *viz.,* seed soaking for 8 h in endosulfan 0.07% + $CaCl_2$ (2%) + one spray with neem leaf extract 5% + whorl application of carbofuran 3G @ 7.5 kg/ha at 30 days after emergence (DAE) recorded significantly lower deadhearts due to shoot fly (21.6%) and higher grain yield (23.3 q/ha) irrespective of planting times. The above main treatment was at par with seed soaking for 8 h in endosulfan 0.07% + $CaCl_2$ (2%) + whorl application of carbofuran 3G @ 7.5 kg/ha at 30 DAE with grain yield of 22.9 q/ha irrespective of planting times.

INTRODUCTION

Sorghum (*Sorghum bicolor* (L) Moench) is an important food and fodder crop in the semi-arid tropics. More than 150 insect pest species have been reported on sorghum (Reddy & Devies, 1979) of which Sorghum shoot fly, *Atherigona soccata* Rondani is a serious pest particularly in late sown crop. In sorghum nearly 32.1 per cent of the actual produce is lost every year due to insect pests (Borad & Mittal, 1983). Granular systemic insecticides through soil application and seed treatment were found effective but they are not only uneconomical but hazardous also. The neem is reported to contain different components with antifeedant properties (Mansour *et al.*, 1986). As sorghum is low economic return crop and grown by the poor farmers, there is a need to develop low cost technologies. Therefore, in order to reduce the cost of application of pesticides, various components of integrated pest management *viz*, sowing time, seed rate, botanical and chemical insecticides were tested against shoot fly in *rabi* sorghum.

METERIAL & METHODS

A field plot experiment was carried out during *rabi* 2001-02 at the Regional Agricultural Research Station, Bijapur in a factorial randomized block design with two factors, eight treatment combinations & three replications. The main factors included (1) Early Planting with normal seed rate (8 kg/ha) and (2) Normal Planting with high seed rate (10 kg/ha) and thinning at 20 days after emergence (DAE). The sub treatments included (1) Seed soaking for 8 h in endosulfan 0.07% +$CaCl_2$ (2%) +One spray with neem leaf extract 5% + whorl application of carbofuran 3G @ 7.5 kg/ha at 30 DAE (2) Seed soaking for 8hrs in endosulfan 0.07% + $CaCl_2$ (2%) + whorl application of carbofuran 3G @ 7.5 kg/ha at 30 DAE (3) Whorl application of carbofuran 3G @ 7.5 kg/ha at 30 DAE and (4) Untreated check. The plot size was 3.6 x 4.5m, each having 6 rows of 4.5m lengths. Sorghum hybrid CSH 19R was used as a test crop. The crop was protected from all pests and diseases by following the recommended plant protection schedule except the shoot fly control. The shoot fly incidence was recorded on 28th day after the emergence of the crop. Total number of plants and number of plants showing deadheart symptoms were recorded in each treatment on 28th day after emergence of the crop. The percentage of deadhearts caused by shoot fly was worked out and subjected to angular transformations before statistical analysis. Later, grain yield was also recorded & converted to per hectare and analyzed statistically.

RESULTS & DISCUSSION

The data recorded on per cent shoot fly deadhearts and grain yield are presented in Table 1. Early planting with normal seed rate

(8 kg/ha) proved to be significantly superior over late planting with high seed rate (10 kg/ha) with respect to reduction in shoot fly incidence and harnessing higher seed yield irrespective of main treatments. The main treatment *viz.,* T_1 (seed soaking for 8 h in endosulfan 0.07% + $CaCl_2$ (2%) + one spray with neem leaf extract 5% + whorl application of carbofuran 3G @ 7.5 kg/ha at 30 DAE) recorded significantly lower deadhearts due to shoot fly (21.6%) irrespective of planting time. The main treatments *viz.,* T_1 (seed soaking for 8 h in endosulfan 0.07% + $CaCl_2$ (2%) + one spray with neem leaf extract 5% + whorl application of carbofuran 3G @ 7.5 kg/ha at 30 DAE) and T_2 (seed soaking for 8 h in endosulfan 0.07% + $CaCl_2$ (2%) + whorl application of carbofuran 3G @ 7.5 kg/ha at 30 DAE) recorded higher grain yield of 23.3 and 22.9 q/ha respectively irrespective of planting time. These two main treatments were at par with each other with respect to grain yield. The interaction effects were non-significant. However, the treatment combination having early planting with normal seed rate (8 kg/ha) along with seed soaking for 8 h in endosulfan 0.07% + $CaCl_2$ (2%) + one spray with neem leaf extract 5% + whorl application of carbofuran 3G @ 7.5 kg/ha at 30 DAE recorded lowest shoot fly incidence (20.8%) and highest grain yield (23.9 q/ha). The results of the present study are in conformity with the results of Men *et al.,* (1985), Balikai *et al.,* (1998), Kandalkar *et al.,* (1994 & 2001) and Ameta (2004).

Table 1. Effect of different IPM modules on the shoot fly incidence and grain yield

Sub Treatments / Main Treatments	% DH due to shoot fly			Grain Yield (q/ha)		
	EP	NP	Mean	EP	NP	Mean
1) Seed soaking for 8hrs in endosulfan 0.07% + $CaCl_2$ (2%) + One spray with neem leaf extract 5% + whorl application of carbofuran 3G @ 7.5 kg/ha at 30 DAE	20.8 (27.1)*	22.4 (28.3)	21.6 (27.7)	23.9	22.6	23.3
2) Seed soaking for 8hrs in endosulfan 0.07% + $CaCl_2$ (2%) + whorl application of carbofuran 3G @ 7.5 kg/ha at 30 DAE	26.5 (31.0)	28.7 (32.4)	27.6 (31.7)	23.3	22.5	22.9
3) Whorl application of carbofuran 3G @ 7.5 kg/ha at 30 DAE	38.7 (38.5)	43.3 (41.2)	41.0 (39.8)	17.8	16.6	17.2
4) Untreated check	39.3 (38.8)	42.4 (40.6)	40.9 (39.7)	15.3	14.3	14.8
Mean	31.3 (33.9)	34.2 (35.6)	32.8 (34.7)	20.1	19.0	19.6

Contd...

Table 1. Contd...

Deadhearts			Grain Yield	
Source	S.Em.±	C.D. (0.05)	S.Em.±	C.D. (0.05)
Main treatments	0.2	0.6	0.2	0.6
Planting dates	0.3	0.8	0.3	0.8
Interaction	0.4	NS	0.4	NS

EP= Early Planting with normal seed rate (8 kg/ha)

NP= Normal Planting with high seed rate (10 kg/ha) and thinning at 20 days after emergence (DAE)

DH= Deadhearts

*Arc sin transformations

CONCLUSIONS

Early planting of sorghum with normal seed rate (8 kg/ha) proved to be significantly superior over late planting with high seed rate (10 kg/ha) with respect to reduction in shoot fly incidence on sorghum and harnessing higher seed yield irrespective of main treatments. The main treatment *viz.*, seed soaking for 8 h in endosulfan 0.07% + $CaCl_2$ (2%) + one spray with neem leaf extract 5% + whorl application of carbofuran 3G @ 7.5 kg/ha at 30 days after emergence (DAE) recorded significantly lower deadhearts due to shoot fly (21.6%) and higher grain yield (23.3 q/ha) irrespective of planting times. The above main treatment was at par with seed soaking for 8 h in endosulfan 0.07% + $CaCl_2$ (2%) + whorl application of carbofuran 3G @ 7.5 kg/ha at 30 DAE with grain yield of 22.9 q/ha irrespective of planting times.

References

Ameta, O. P. 2004. Management of Shoot fly, *Atherigona soccata* and stem borer, *Chilo partellus* in sorghum. *Indian J. Appl. Ent.* 18(1):87-89.

Balikai, R. A., A. P. Biradar. & R. C. Kolhar. 1998. Management of sorghum shoot fly, *Atherigona soccata* Rondani. *Adv. Agric. Res. India.* 9:47-49.

Borad, P. K. & V. P. Mittal. 1983. Assessment of losses caused by pest complex of sorghum hybrid, CSH-5. PP. 271-278. In : *Crop losses due to insect pests.* (Eds.) B.H. Krishnamurthy Rao & K.S.R.K.Murthy. Entomological Society of India, Hyderabad.

Kandalkar, H. G., U. B. Men, S. S. Narkhed & V. B. Shekhar. 1994. Efficacy of some plant products against sorghum shoot fly, *Atherigona soccata* Rondani. *PKV Res. J.* 18(2):186-189.

Kandalkar, H. G., K. J. Morey, U. B. Men & S. B. Atale. 2001. Integrated Pest Management for shoot fly, *Atherigona soccata* Rondani. *Pestology* 25(6):45-47.

Mansour, F. A., K. R. S. Ascher & N. Omari. 1986. Toxicity of neem (*Azadirchta india*) seed kernel extracts on the spider, *Chiracanthium mildei*. *Phytoparasitica*. 14(1):73-76.

Men, U. B., S. K. Mundiwale. & M. N. Borle. 1985. Chemical control of sorghum shoot fly. *Sorghum Newsletter* 28: 53.

Reddy, K. V. S. & J. C. Devies. 1979. Pests of sorghum and pearl millet and their parasites and predators recorded at ICRISAT centre. *Central Entomology Progress Report No. 2,* International Crop Research Iinstitute for Semi Arid Tropics, Patancheru, Andhra Pradesh.

49

Mechanical Control of Red-headed Hairy Caterpillar, *Amsacta albistriga* in Groundnut

GANIGER, P.C. AND V.T. SANNAVEERAPPANAVAR

ABSTRACT

A study was taken up to assess the cost of manual collection of grown up larvae (sixth and seventh instars) in groundnut crop by engaging men and women labourers. Each labourer collected 115.70 larvae in 100 m^2 area of the crop in 30 minutes. At this rate, a labourer would collect all the larvae in one hectare of the crop in about 22 hours. The cost of manual collection of larvae in one hectare would be Rs. 110.00 at the prevailing daily wage of Rs. 40.00 per day per person in Pavagada area. This method apart from being safer, is more economical compared to the cost involved in chemical (Fenvalerate 0.4 %) suppression of the pest (Rs. 250/ha).

INTRODUCTION

The Red-headed hairy caterpillar (RHC), *Amsacta albistriga* (Walker) (Lepidoptera: Arctiidae) is one of the major pests causing severe damage to several crops in South India (Ghewande & Nandagopal, 1997). In Karnataka, *A. albistriga* has been a major pest in red loamy soils in the districts of Raichur, Bellary, Gulbarga,

Belgaum, Bijapur, Chitradurga, Chikkamagalur and Kolar (Puttarudriah, 1956 ; Thontadarya *et al.*, 1976). Recently, an outbreak was also reported from some parts of Bidar and Bellary districts. In endemic areas, groundnut and other crops such as sesamum, cowpea, green gram, sunflower, etc. suffer heavy defoliation from large populations of RHC (Vinayakachari, 1996) and the farmers often resort to resowing of crops. Mechanical control provides simple, effective and economic way of managing the pest insect.

MATERIAL AND METHODS

One of the methods recommended for suppressing this pest is manual collection and destruction of egg masses and larvae. A study was taken up to assess the cost of manual collection of grown up larvae (sixth and seventh instars) in groundnut crop by engaging men and women labourers. The cost of mechanical control of RHC in one hectare was estimated based on the time taken by each labourer to collect the larvae in 100 m^2 area.

RESULTS & DISCUSSION

Each labourer collected 115.70 larvae in 100 m^2 area of the crop in 30 minutes. At this rate, a labourer would collect all the larvae in one hectare of the crop in about 22 hours. The cost of manual collection of larvae in one hectare would be Rs. 110.00 at the prevailing daily wage of Rs. 40.00 per day per person in Pavagada area. This method apart from being safer, is more economical compared to the cost involved in chemical (Fenvalerate 0.4 %) suppression of the pest (Rs. 250/ha). The mechanical collection and destruction of egg masses as well as the first instar and grown up caterpillars, and pupae, was also found to be useful in reducing the pest population (Nagarajan *et al.*, (1957).

Table 1. Cost of manual collection of grown up larvae of *Amsacta albistriga*

No. of larvae* collected/person in100m²/30 min	Hrs. req/ha	Labour cost/ ha (Rs.)	Cost of chemical Supplession/ha (Rs)
115.70	22.00	110.00	250.00

* Mean of 10 replications

* * Daily wages at Rs. 40.00 per person

References

Ghewande, M. P. & V. Nandagopal. 1997. Integrated pest management in groundnut (*Arachis hypogea* L.) in India. *Integrated Pest Mang. Reviews.* 2: 1-15.

Nagarajan, K. R., K. Perumal & N. Shanmugam. 1957. The red-headed hairy caterpillar, *Amsacta albistriga* W. and its field-scale control. *Madras Agric. J.* 44 (4): 150-53.

Puttarudriah, M. 1956. Pests of ragi and their control (including Bajra, Tenai, *Wheat and Maize* (in Kannada). Dept. Agri., Mysore Pl. Prot. Booklet No.3.23pp.

Thontadarya, T. S., M. C. Devaiah, Puttaswamy & R. Govindan. 1976. Outbreak of red headed hairy caterpillar, *Amsacta albistriga* Walker on groundnut in Ramdurga area (Belgaum District, Karnataka). *Curr. Res.* 5: 175-176.

Vinayakachari, M. L. 1996. *Systematic and biological studies on arctiid hairy caterpillar pests (Lepidoptera: Arctiidae) with special reference to red-headed hairy caterpillar, Amsacta albistriga (Walker).* M. Sc Thesis submitted to University of Agricultural Sciences, Bangalore.

50

Bio-efficacy of Ready-mix Formulations of Insecticides Against Pigeon Pea Pod Borer Complex

GANIGER, P. C., D. W. WADNERKAR
AND M.THIPPAIAH

ABSTRACT

A field experiment was conducted to test the efficacy of readymix formulations of insecticides viz., chloropyriphos 50 EC + cypermethrin 5 EC, triazophos 35 EC + deltamethrin 1 EC, Profenophos 40 EC + cypermethrin 4 EC along with three individual insecticides i.e. cypermethrin 10 EC, Quinalphos 35 EC, Monocrotophos 36 SL against pigeonpea pod borer complex during 1999-2000. All the readymix formulations of insecticides tried significantly reduced pod borer damage over untreated control. The combination of chlorpyriphos + cypermethrin 500 + 50 g. a. i./ha recorded lowest pod damage (23.43 %) and grain damage (10.56 %) due to both *Helicoverpa armigera* and *Exelastis atomosa.* Next treatments viz., chlorpyriphos + cypermethrin 400+ 40 g. a. i./ha (32.73 and 12.28 %) and triazophos + deltamethrin 350 + 10 g. a. i./ha (31.28 and 13.16 %) were found superior. The infestation due to pod fly, *Melanoagromyza obtuse* was 6.10 per cent in triazophos in deltamethrin 350 + 10 g. a. i./ha. However minimum per cent grain damage of 1.69 was seen in the treatment, chlorpyriphos + cypermethrin 500 + 50 g. a. i./ha with highest yield of 19.94 q/ha.

INTRODUCTION

Pigeon pea is an important legume crop grown in all the states of India. Gram pod borer, *Helicoverpa armigera* (Hub.), tur plume moth, *Exelastis atomosa* Walsh and pod fly, *Melanoagromyza obtusa* Mall. are the key pests, collectively referred to as pod borer complex of pigeon pea. Awasthi & Bhatnagar (1983) reported pod and grain damage due to pod borer complex to the extent of 77.04 and 68.70 per cent, respectively. Different pesticide firms have developed a number of products to manage these pests in India. These are reported to be advantageous when several insects attack the crop simultaneously. Also the combination products are considered as oldest tools to manage delayed evolution of resistance in insects (Leeper *et al.*, 1986).

A ready-mix formulation of chlorpyriphos 50 EC + cypermethrin 5 EC (Nurelle-D- 505) developed by M/s DE-NOCIL Crop protection Ltd., Mumbai was evaluated for its efficacy against pod borer complex in pigeon pea.

MATERIAL AND METHODS

Pods with large circular holes and grains completely devoured by the caterpillar, were considered as damaged by *H. armigera*. The holes made by the larvae of plume moth were small or medium. The holes were opposite to the spot where seed is attached to the pod (inside) and partially eaten grains with blackish excreta. The pods with externally visible brownish pin holes and showing mines or gnawing on the seeds were considered as damaged due to pod fly, *M. obtusa*. Randomly selected 100 pods at harvest were observed for the number of seeds damaged by *H. armigera, E. atomosa* and *M. obtusa*. The data obtained in percentage were angular transformed and statistically analysed. The yields were converted to q/ha and analysed.

A field experiment was conducted during *kharif* 1999-2000 at College of Agriculture Farm, Marathawada Agricultural University, Parbhani (Maharashtra), to study the efficacy of Nurelle-D-505 (chlorpyrifos + cypermethrin) in comparison with other recommended insecticides against pigeon pea pod borers. A sterility mosaic resistant variety BSMR-736 was sown in plot size of 4.8 X 3.9 m with spacing of 60 cm X 30 cm and the experiment was laid in a randomized block design with three replications. Three ready mix formulations *viz.*, chlorpyrifos 50 EC + cypermethrin 5 EC (300+30,400+40,500+50 g. a.

i. /ha), triazophos 35 EC + deltamethrin 10 EC (350 + 10 g.a.i./ha) and four insecticides *viz.*, chlorpyrifos, cypermethrin, quinalphos, monocrotophos were used individually. Spraying was initiated at bud initiation and repeated there after at 15 days interval. During cropping period, 3 sprays were given. The insecticides were applied by using foot sprayer (pedal pump).

RESULTS & DISCUSSION

Nurelle-D-505, the combination product of chlorpyrifos and cypermethrin, has quick knock down effect coupled with long persistence. It is formulated as emulsifiable concentrate (55 percent active ingredient).

Pod borer damage: The lowest per cent pod damage by *H. armigera* and *E. atomosa* was observed in chlorpyriphos + cypermethrin 550 g. a.i./ha (23.43 %) followed by triazophos + deltamethrin 360 g a. i./ha (31.28 %) and chlorpyriphos +cypermethrin 330 g.a.i. /ha (31.86 %).

The per cent infestation of grains varied from 10.56 to 23.09. Chlorpyriphos + cypermethrin 550 g. a.i./ha recorded lowest grain damage (10.56 per cent) and was significantly superior over other treatments. No phytotoxicity symptoms such as yellowing, epinasty, hyponasty etc were observed in pigeon pea, sprayed with ready mix formulations of different concentrations.

Chhabra *et al.*, (1980) reported that synthetic pyrethroids were most effective for the control of lepidopterous and dipterous pests. Singh *et al.*, (1988) proved the efficacy of monocrotophos and cypermethrin against lepidopteran borers. Deltamethrin, cypermethrin resulted in highest yield. Tambe *et al.*, (1997) reported triazophos + deltamethrin @ 350 + 10 g. a. i./ha as effective in reducing per cent infestation due to pigeon pea pod borers and also recorded highest yield.

Pod fly damage: The per cent pod damage due to pod fly (*M. obtuse*) varied from 6.10 to 13.97. The lowest 6.10 per cent pod damage was observed in triazophos + deltamethrin 660 g.a. i./ha followed by profenofos + cypermethrin 360 g.a. i./ha (6.39). However, per cent grain damage was minimum in chlorpyriphos + cypermethrin 440 g. a. i./ha (1.69) followed by triazophos + deltamenthrin 360 g.a. i/ha (2.16) and profenofos + cypermethrin 660 g. a. i./ha (2.36).

Babusundra & Rajasekaran (1984) reported 0.07 % triazophos as most effective against pod fly and recorded maximum yield. In the present experiment triazophos 35 EC + deltamethrin 1 EC was found effective in reducing damage by pod fly.

Cypermethrin 0.0125 % was reported to be effective against pod fly by Patel *et al.*, (1991). In the present experiment ready-mix formulation of chlorpyrifos + cypermethrin was found to be effective against grain damage caused by pod fly.

Grain yield: The maximum grain yield was recorded in the treatments with chlorpyrifos + cypermethrin 550 g. a. i./ha (19.94 q/ ha) and 440 g. a. i./ha (18.49 q/ha).

Effectiveness of several combination products including chlorpyrifos + cypermethrin has been proved against cotton bollworms (Shrinivasan *et al.*, 1995) and fruit borers of *bendi* (Girradi *et al.*, 1998). In the present investigation the effectiveness is realized against pod borer complex in pigeon pea also.

ACKNOWLEDGEMENTS

The authors are thankful to DE-NOCIL Crop protection, Mumbai for providing ready-mix formulations for the conduct of this experiment.

References

Awasthi, J. K. & Bhatnagar, A. 1983. A note on damage caused by pod borer complex in pigeon pea. *Bull. Entom.* 24 (1): 37 – 40.

Babusundra, P. C. & B. Rajesekaran. 1984. Studies on the control of pod borer and pod fly on red gram (*Cajanus cajan* L.). *Pesticides.* 18 (2): 24-26.

Chhabra, K. S., B. S. Kouner. & J. S. Brar. 1980. Synthetic pyrethroids for control of pigeon pea pest complex. *Proceedings of the International Workshop on Pigeon pea.* 15-19 Dec., 1980. Internatonal Centre for Research in Semi Arid Tropics, Patancheru, Andhra Pradesh, India. Pp. 349-354.

Girradi, R. S., S. Lingappa. & R. Heggade. 1998. Bio-efficacy of chlorpyrifos (Dursban) + cypermethrin (Cilcord) against insect pests of Bhendi. *Pestology* XII (6): 25-28.

Leeper, J. R., Roush & H. T. Reynolds. 1986. Preventing or managing resistance in arthropods. In: *Pesticide resistance; strategies and tactics for management.* National Academy of Sciences, Washington D. C. pp. 335-346.

Patel, V. J., C. B. Patel, R. C. Jhala. & A. B. Rai. 1991. Comparative efficacy of some synthetic pyrethroids against pod flies on pigeon pea. *Pestology* XII(1): 19-23.

Singh, H. M., R. Singh. & S. M. Rizvi. 1998. Insecticidal evaluation against pigeon pea pod borers. *Indian J. P. Prot.* 16: 217-220.

Shrinivasan, M. R., V. Sardana. & G. Krishnamurthy. 1995. Field evaluation of a combination product (chlorpyrifos 22.5 + cypermethrin 2.5 EC) and aquaflowable formulations against cotton pests. *Pestology* XIX (13): 31-34.

Tambe, A. B., U.K. Kadam. & K. S. Darekar. 1997. Efficacy of ready-mix formulations of insecticides against pigeon pea pod borer complex. *Pestology* XIX (4): 17-19.

51

Seed Treatment as an Eco-friendly Management Tool for Suppression of Insect Pests in Sorghum

BALIKAI, R. A.

ABSTRACT

A field trial was conducted at the Regional Agricultural Research Station, Bijapur, Karnataka during *rabi* season of 2005-06 to find out eco-friendly management tactics for the suppression of insect pests in sorghum. The results revealed that, the seed treatment with thiamethoxam 70WS @ 3 g/kg seeds proved highly effective against shoot fly and significantly superior over rest of the treatments by recording 5.2 per cent dead hearts. The next best treatment in respect of shoot fly suppression was seed treatment with thiamethoxam 70WS @ 2 g/kg seeds which in turn was on par with imidacloprid 70 WS @ 5g/kg seeds. With respect to aphid incidence, three treatments *viz.*, whorl application of carbofuran 3G @ 8kg/ha at 35-40 days after sowing (DAS), spray with endosulfan 35 EC @ 0.07% at 35-40 DAS and seed treatment with thiamethoxam 70WS @ 3 g/kg seeds resulted in low aphid incidence of 9.3, 10.2 and 10.2 per cent aphid index respectively and were highly effective and significantly superior over rest of the treatments. The seed treatment with thiamethoxam 70WS @ 3 g/kg seeds proved highly effective in reducing the disease incidence (5.2%) and significantly superior over rest of the treatments except seed treatment with imidacloprid 70 WS @ 5g/kg seeds (8.2%). Seed treatment with thiamethoxam 70WS @ 3 g/kg seeds recorded highest grain yield (22.5 q/ha) and fodder yield (54.3 q/ha) followed by seed treatment by imidacloprid 70 WS @ 5g/kg seeds (21.7 and 52.1 q/ha grain and fodder yield respectively) and thiamethoxam 70WS @ 2 g/kg seeds (20.49 and 51.3 q/ha grain and fodder yield respectively) which were on par with each other.

INTRODUCTION

Sorghum is vulnerable to over 150 insect species from sowing to the final stage of crop harvest (Sharma, 1985) of which shoot fly, shoot bug and aphid are the important regular pests. The Sorghum shoot fly, *Atherigona soccata* Rondani (Diptera: Muscidae) causes severe damage in the early stage and lasts up to four weeks causing severe reduction in plant population. Maximum yield losses of 75.6% in grain and 68.6% in fodder have been reported by Pawar *et al.*, (1984). Its incidence is greater in late sown crop in rainy and post rainy seasons in India. Several workers have tried different insecticides for this pest (Shivpuje & Thombare, 1983; Patil *et al.*, 1992; Panchabhavi & Kotikal, 1992). Sorghum aphid, *Melanaphis sacchari* (Zehntner) is distributed in Asia, Africa and Americas. It prefers to feed on the under surface of older leaves, resulting in premature drying of leaves, non-filling of grains and deterioration of fodder quality. Spraying or dusting of several insecticides are being recommended for its control. The shoot bug, *Peregrinus maidis* (Ashmead) previously considered to be of minor importance, of late has become a serious pest with the introduction of new sorghum genotypes of different maturity periods in certain parts of Maharashtra, Andhra Pradesh, Karnataka and Tamil Nadu. The shoot bug is a major hurdle in *rabi* sorghum production by causing dual problem of direct loss by sucking the sap and indirect damage by transmitting sorghum stripe virus disease. Hence, it affects yield potential of grain and fodder. Managing the pest in established sorghum ecosystem through chemical spraying has several limitations. Farmers are unable to go for spraying due to increased cost of production of sorghum and also phytotoxic effect of these insecticides on foliage. Hence, different seed dressers were evaluated in comparison with spray and granules for the management of sorghum pests.

MATERIAL & METHODS

A field trial was conducted at the Regional Agricultural Research Station, Bijapur during *rabi* season of 2005-06 under rain fed conditions

Table 1: Effect of insecticides on pest incidence and yield of *rabi* sorghum

S. No.	Treatments	% Dead-hearts due to Shoot fly	% Aphid Index	% Sorghum stripe disease Incidence	Grain Yield (q/ha)	Fodder Yield (q/ha)
1.	Phorate 10 G @ 15 kg/ha (0.9g/m.r.) (Soil application)	20.4 (26.8) e	48.3 (44.0) d	25.2 (30.1) e	16.2 cd	45.1 c
2.	Carbofuran 3G @ 20kg/ha (1.2g/m.r.) (Soil application)	18.2 (25.3) e	47.4 (43.5) d	27.8 (31.8) e	15.1 cd	44.3 c
3.	Imidacloprid 70 WS @ 2g/kg seeds (Seed treatment)	12.6 (20.8) cd	16.5 (24.0) c	14.3 (22.2) cd	18.4 bc	47.2 bc
4.	Imidacloprid 70 WS @ 5g/kg seeds (Seed treatment)	10.5 (18.9) bc	14.3 (22.2) bc	8.2 (16.5) ab	21.7 ab	52.1 a
5.	Thiamethoxam 70 WS @ 2g/kg seeds (Seed treatment)	8.4 (16.8) b	14.3 (22.2) bc	10.5 (18.9) bc	20.4 ab	51.3 ab
6.	Thiamethoxam 70 WS @ 3g/kg seeds (Seed treatment)	5.2 (13.1) a	10.2 (18.6) a	5.2 (13.1) a	22.5 a	54.3 a
7.	Endosulfan 35 EC @ 5.0 ml +20 ml water (Seed treatment)	10.1 (18.5) bc	48.3 (44.0) d	29.3 (32.8) ef	16.5 cd	45.5 c
8.	Imidacloprid 200SL @ 2.0 ml +20 ml water (Seed treatment)	14.2 (22.1) d	45.7 (42.5) d	30.1 (33.3) ef	15.3 cd	45.0 c
9.	Carbofuran 3G @ 8kg/ha (Whorl appln.) at 35-40 DAS	38.5 (38.4) f	10.2 (18.6) a	17.1 (24.4) d	16.2 cd	46.2 c
10.	Endosulfan 35 EC @ 0.07% (Spray) at 35-40 DAS	39.8 (39.2) f	9.3 (17.7) a	16.2 (23.7) d	15.5 cd	44.5 c
11.	Untreated Check	38.7 (38.5) f	50.3 (45.2) d	35.2 (36.3) f	13.3 d	37.4 d
	S.Em.±	1.0	1.5	1.3	1.3	1.5
	C.D. (0.05)	2.9	4.4	3.8	3.7	4.5
	C.V. (%)	6.6	8.4	8.7	12.6	5.7

* Figures in the parentheses indicate arcsin transformations
Figures in the columns followed by same letters are not significantly different at 5% level.
m.r = meter row

in randomized block design replicated thrice with eleven treatments including an untreated check (Table 1). M 35-1 sorghum variety was planted by dibbling the seeds at a spacing of 60x15 cm in a plot size of 3.6x4.5m, each having six rows of 4.5m length. The thinning operation was done a week after emergence of the crop. For granular insecticides in T_1 and T_2, granules were applied in seed furrow at the time of sowing. For the treatments involving seed treatment (from T_3 to T_6), sorghum seeds were treated with test insecticide using 1 per cent acacia gum followed by shade drying for three hours before sowing. For the treatments involving seed treatment (T_7 and T_8), sorghum seeds were treated with test insecticides using 20 ml of water per kilo gram of seeds, followed by shade drying for three hours before sowing. The total number of plants and number of plants showing deadheart symptoms were recorded in each treatment on 28th day after emergence of the crop. The percentage of dead hearts caused by shoot fly was worked out and subjected to angular transformations before analysis. The per cent aphid index was recorded by following the methods of Balikai & Lingappa (2002) when the aphid incidence was at its peak during January second week. Total number of plants and number of plants showing sorghum stripe disease (SSD) symptoms were recorded from the net plot of each treatment. The data were subjected to angular transformations before statistical analysis. Later, grain and fodder yields were recorded, converted to quintals per hectare and analyzed statistically.

RESULTS

Among the eleven treatments, the maximum incidence of shoot fly (38.7% dead hearts) was recorded in the untreated plot (Table 1). All the insecticidal treatments significantly reduced the incidence of shoot fly except whorl application of carbofuran 3G @ 8kg/ha at 35-40 days after sowing (DAS) and spray with endosulfan 35 EC @ 0.07% at 35-40 DAS. The seed treatment with thiamethoxam 70WS @ 3 g/kg seeds proved highly effective against shoot fly and significantly superior over rest of the treatments by recording 5.2 per cent dead hearts. The next best treatment in respect of shoot fly suppression was seed treatment with thiamethoxam 70WS @ 2 g/kg seeds which in turn was on par with imidacloprid 70 WS @ 5g/kg seeds.

With respect to aphid incidence, three treatments viz., whorl application of carbofuran 3G @ 8kg/ha at 35-40 days after sowing (DAS), spray with endosulfan 35 EC @ 0.07% at 35-40 DAS and seed treatment with thiamethoxam 70WS @ 3 g/kg seeds with low aphid incidence of 9.3, 10.2 and 10.2 per cent aphid index, respectively were highly effective and significantly superior over rest of the treatments. The next best treatments in respect of aphid suppression included seed treatment with thiamethoxam 70WS @ 2 g/kg seeds, seed treatment with imidacloprid 70 WS @ 5g/kg seeds and seed treatment with imidacloprid 70 WS @ 2g/kg seeds which recorded 14.3, 14.3 and 16.5 per cent aphid index, respectively. The remaining treatments (soil application of Phorate 10 G @ 15 kg/ha (0.9g/m.r.), soil application of Carbofuran 3G @ 20kg/ha (1.2g/m.r.), seed treatment with Endosulfan 35 EC @ 5.0 ml + 20 ml water and seed treatment with Imidacloprid 200SL @ 2.0 ml + 20 ml water) were significantly inferior to all the above treatments in giving protection against aphids.

As far as the disease incidence is concerned, all the insecticidal treatments significantly reduced the disease incidence except seed treatment with Endosulfan 35 EC @ 5.0 ml + 20 ml water and seed treatment with Imidacloprid 200SL @ 2.0 ml + 20 ml water. The seed treatment with thiamethoxam 70WS @ 3 g/kg seeds proved highly effective in reducing the sorghum stripe disease incidence (5.2%) and significantly superior over rest of the treatments except seed treatment with imidacloprid 70 WS @ 5g/kg seeds (8.2%). The next best treatment in respect of disease suppression was seed treatment with thiamethoxam 70WS @ 2 g/kg seeds (10.5%) which in turn was on par with imidacloprid 70 WS @ 2g/kg seeds (14.3%).

With regard to grain yield, seed treatment with thiamethoxam 70WS @ 3 g/kg seeds recorded highest grain yield (22.5 q/ha) and it was on par with seed treatment by imidacloprid 70 WS @ 5g/kg seeds (21.7 q/ha) and thiamethoxam 70WS @ 2 g/kg seeds (20.49 q/ha). The next best treatment in this respect was seed treatment with imidacloprid 70 WS @ 2g/kg seeds (18.4 q/ha) which was on par with all other treatments except seed treatment with thiamethoxam 70WS @ 3 g/kg seeds and untreated check. With respect to fodder yield, seed treatment with thiamethoxam 70WS @ 3 g/kg seeds recorded highest fodder yield (54.3 q/ha) followed by seed treatment with imidacloprid 70 WS @ 5g/kg seeds (52.1 q/ha) and thiamethoxam 70WS @ 2 g/kg seeds (51.3 q/ha) and were on par with each other. The latter treatment was on par with seed treatment by imidacloprid 70 WS @ 2g/kg seeds (47.2 q/ha) which in turn was on par with all other remaining treatments except untreated check.

DISCUSSION

The seed treatment with thiamethoxam 70WS @ 3 g/kg seeds proved highly effective against shoot fly and significantly superior over rest of the treatments by recording 5.2 per cent dead hearts followed by seed treatment with thiamethoxam 70WS @ 2 g/kg seeds which in turn was on par with imidacloprid 70 WS @ 5g/kg seeds. Karibasavaraja *et al.,* (2005) also reported that, the seed treatment with thiamethoxam 70 WS @ 5 and 4 g. a.i./kg seeds were very effective in reducing shoot fly incidence even under high pest population. The results of the present study are in conformity with the findings of Balikai (2003) who reported 4.5 per cent shoot fly dead hearts with thiamethoxam 70 WS @ 4 g/kg seeds. The results are also in agreement with the findings of Mote *et al.,* (1995), Hiremath *et al.,* (1995) and Balikai (1998) with respect to efficacy of imidacloprid. Similarly, the results of study conducted by Katole *et al.,* (2003) also support the present findings. With respect to aphid incidence, the results are in agreement with the findings of Balikai & Lingappa (2003) with respect to efficacy of imidacloprid 70 WS. The seed treatment with thiamethoxam 70WS @ 3 g/kg seeds proved highly effective in reducing the sorghum stripe disease incidence (5.2%) and significantly superior over rest of the treatments except seed treatment with imidacloprid 70 WS @ 5g/kg seeds (8.2%). The next best treatment in respect of disease suppression was seed treatment with thiamethoxam 70WS @ 2 g/kg seeds (10.5%) which in turn was on par with imidacloprid 70 WS @ 2g/kg seeds (14.3%). With regard to grain yield, seed treatment with thiamethoxam 70WS @ 3 g/kg seeds recorded highest grain yield (22.5 q/ha) and it was on par with seed treatment by imidacloprid 70 WS @ 5g/kg seeds (21.7 q/ha) and thiamethoxam 70WS @ 2 g/kg seeds (20.49 q/ha). These results corroborate with the findings of previous workers with respect to imidacloprid (Mote *et al.,* 1995; Hiremath *et al.,* 1995; Balikai, 1998; Bheemanna *et al.,* 2003). The results of the present study are in conformity with the findings of Balikai (2003) who reported highest grain yield of 32.1q/ha with thiamethoxam 70 WS @ 4 g/kg seeds. Similarly, the results of study conducted by Katole *et al.,*(2003) also support the present findings.

CONCLUSION

The seed treatment with thiamethoxam 70WS @ 3 g/kg seeds proved highly effective against shoot fly and significantly superior over rest of the treatments by recording 5.2 per cent dead hearts. The next best treatment in respect of shoot fly suppression was seed treatment with thiamethoxam 70WS @ 2 g/kg seeds which in turn was on par with imidacloprid 70 WS @ 5g/kg seeds. With respect to aphid incidence, three treatments viz., whorl application of carbofuran 3G @ 8kg/ha at 35-40 days after sowing (DAS), spray with endosulfan

35 EC @ 0.07% at 35-40 DAS and seed treatment with thiamethoxam 70WS @ 3 g/kg seeds with low aphid incidence of 9.3, 10.2 and 10.2 per cent aphid index, respectively were highly effective and significantly superior over rest of the treatments. The seed treatment with thiamethoxam 70WS @ 3 g/kg seeds proved highly effective in reducing the sorghum stripe disease incidence (5.2%) and significantly superior over rest of the treatments except seed treatment with imidacloprid 70 WS @ 5g/kg seeds (8.2%). Seed treatment with thiamethoxam 70WS @ 3 g/kg seeds recorded highest grain yield (22.5 q/ha) and fodder yield (54.3 q/ha) and was followed by seed treatment by imidacloprid 70 WS @ 5g/kg seeds (21.7 and 52.1 q/ha grain and fodder yield respectively) and thiamethoxam 70WS @ 2 g/kg seeds (20.49 and 51.3 q/ha grain and fodder yield respectively) which were on par with each other.

References

Balikai, R. A. 1998. Evaluation of imidacloprid 70WS against Sorghum Shoot Fly. *Agric. Sci. Digest* 18: 85-87.

Balikai, R. A. 2003. Evaluation of new insecticides for the management of shoot fly, *Atherigona soccata* (Rondani) on *rabi* sorghum. *National Symposium on Biodiversity Management for 21st Century,* Institute of Agricultural Sciences, Banaras Hindu University, Varanasi, UP (India). June 28-30, 2003. Pp.128-129.

Balikai, R. A. & S. Lingappa. 2002. Avoidable loss in different genotypes of sorghum due to aphid, *Melanaphis sacchari* (Zehntner). *J. Aphidology* 16:125-128.

Balikai, R. A. & S. Lingappa. 2003. Management of aphid, *Melanaphis sacchari* (Zehntner) on *rabi* sorghum. *J. Aphidology* 17(1&2): 67-72.

Bheemanna, M., B. V. Patil, S. G. Hanchinal, V. Anandkumar & Suresh. 2003. Management of sorghum shoot bug by seed treatment. *National Symposium on Frontier Areas of Entomological Research,* Indian Agricultural Research Institute, New Delhi. November 5-7, 2003. 25pp

Hiremath, I. G. S. G. Bhuti & S. Lingappa. 1995. Imidacloprid- a new promising seed dress for the management of sorghum shoot fly. *Karnataka J. Agric. Sci.* 8:163-167.

Karibasavaraja, L. R., R. A. Balikai & V. P. Deshpande. 2005. Thiamethoxam 70 WS, A new promising seed dress for the suppression of sorghum shoot fly. *Ann. Pl. Protec. Sci.* 13(1):85-87.

Katole, S. R., N. V. Kudale. & S. M. Thakare. 2003. New systemic insecticides against internal feeders of sorghum. *National Symposium on Frontier Areas of Entomological Research,* Indian Agricultural Research Institute, New Delhi. November 5-7, 2003. pp.26.

Mote U. N., A. P. Mohite & G. R. Lolage. 1995. Seed treatment of imidacloprid against sorghum shoot fly. *Pestology* 19:37-40.

Panchbhavi, K. S. & Y. K. Kotikal. 1992. Effect of seed soaking in insecticide solution on the germination and on the incidence of shoot fly. *Curr. Res.* 21:29-30.

Patil, B.V., M. Bheemanna, Somasekhar & S.C. Hiremath. 1992. Effect of seed soaking of insecticides on sorghum shoot fly, *Atherigona soccata* Rondani. *Indian J. Ent.* 54: 424-427.

Pawar, V. M., G. D. Jadhav & B. S. Kadam. 1984. Compatibility of Oncol 50 SP with different fungicides on sorghum (CS-3541) against shoot fly (*Atherigona soccata* Rondani). *Pesticides* 18:9-10.

Sharma, H. C. 1985. Strategies for pest control in sorghum in India. *Tropical Pest Management* 31:167.

Shivpuje, P. R. & P. A. Thombare. 1983. Efficacy of different insecticides for the control of sorghum shoot fly. *J. Maharashtra Agri. Univ.* 8:83.

52

Eco-friendly Management of Rodent Pests Using Tanjore-bow Trap in Irrigated Rice Fields of Tiruchirappalli District Tamil Nadu

SAKTHIVEL, P., A. BHARATHIRAJA
AND P. NEELANARAYANAN

ABSTRACT

Rodents are deemed to be the most destructive vertebrate pests of agricultural crops. Tanjore-bow trap is an inexpensive and eco-friendly tool for rodent management by farmers in Cauvery delta. In a portion of Cauvery delta of Tiruchirappalli district, field trials using traps to control rodent pests were conducted during two rice crop seasons, *viz., Kuruvai* (July – September 2006) and *Samba* (October 2006- February 2007). During *Kuruvai* season 26 ha rice fields were subjected to the trials for 66 days and during *Samba* season 54 ha rice fields were covered for 79 days. The total number of trap nights during *Kuruvai* and *Samba* seasons' rice crop were 6440 (range/day-80 to 240) and 14746 (range/day -120 to 720), respectively. Three species of field rodents *viz., Bandicota bengalensis, Millardia meltada,* and *Mus booduga* were trapped during both seasons of rice crop. More *B. bengalensis* were trapped than *M. meltada* and *M. booduga* during both seasons. The percent trap success was found to be 15 during *Kuruvai* season and 16 during *Samba* season. The total amount spent by the farmers was Rs. 5162 for *Kuruvai* season and Rs. 12115 for *Samba* season. The cost of trapping a single rodent was computed to be Rs. 5 during both seasons. From the results of the present study, it can be suggested that trapping can be an effective method for controlling rice field rodents. It is also eco-friendly and non-hazardous.

INTRODUCTION

Rodents are by and large the most destructive vertebrate pests on earth. They are also deemed to be the principal foes of farmers since time immemorial. According to Malhi & Sheikher (1989), magnitude of rodent depredation is 10-15% of total national agricultural produce in India. Besides, they spread many diseases among humans and livestock and also destroy many household articles.

Rice is the staple food crop of *Cauvery* Delta, in Tamil Nadu and in addition pulses, plantain, soy bean, sugarcane, cotton, sesame etc., are cultivated. The crop fields are inhabited by the lesser bandicoot rat (*Bandicota bengalenis*), the soft-furred field rat (*Millardia meltada*) and the Indian field mouse (*Mus booduga*) while the Indian Gerbil (*Tatera indica*) inhabits the barren lands adjacent to the crop fields (Neelanarayanan *et al.*, 1996). All these rodent species cause damage to the cultivated crops. Management of these rodent pests is indispensable in places where the problem is acute.

Many rodentologists from different parts of our country have reported the magnitude of rodent depredation in different stages of various crops. The study made by Neelanarayanan (1997), in the *Cauvery* delta, revealed that paddy, pulses (Green gram and Black gram), sugarcane and cotton crops, during pre-harvest stage, suffered up to a maximum of 37%, 58%, 11% and 14% damage, respectively due to rodent pests. It is thus apparent that these vertebrate pests are constant threats to our country's food production and hence management of their population at below economic threshold levels in different crop fields is an urgent need. Our ancestors adapted simple, economic, effective and non-hazardous (to non-targets) control methods against them. However, the advent of rodenticides changed the attitudes of our farmers and chemicals replaced the traditional methods. When man began to use novel methods of rodent control like use of acute and chronic rodenticides and chemical fumigants, he was totally unaware of the dire consequences of these chemicals to his livestock and other non-target organisms.

In U.S.A., brodifacoum (a second generation chronic rodenticide) used for rats, house mice and vole control, affected the non-target species like barn owl, screech owl and other raptors through secondary poisoning (Hegdal & Blaskiewicz, 1984; Hegdal & Colvin, 1988). In India, Neelanarayanan *et al.*, (1994) reported the scope for secondary poisoning hazards of zinc phosphide (an acute rodenticide) to house crow and jungle crow. The indirect potential hazards of zinc phosphide to non-targets animals have been reported by Neelanarayanan & Kanakasabai (1995). In view of the side effects of rodenticides on the non-target animals and environment, scientists are searching for rodent control methods, which will not create any pollution on the ecosystem. To accomplish this objective an attempt has been made to study the efficacy of *Tanjore* bow trap in the crop fields and to contain the population of rodents and the results thereof are presented in this paper.

MATERIAL & METHODS

Tanjore bow-trap is an inexpensive and eco-friendly tool for rodent control but unfortunately many farmers of our country are unaware of it. Field trials of Tanjore bow traps to reduce rodent population was conducted in a portion of Cauvery delta in two rice crop seasons, *viz., Kuruvai* (July – September 2006) and *Samba* (October 2006- February 2007). During *Kuruvai* season 26 ha rice fields were subjected to the trials for 66 days and during *Samba* season 54 ha rice fields were covered for 79 days. Experienced traditional rodent trapper(s) were hired for this purpose. The trapper(s) used (80-720/ day) *Tanjore* bow-traps according to the need and size of the crop fields. The trapper(s) placed the traps in the evening hours in a grid pattern at 2-5 m interval and left them overnight. Parched paddy and raw rice mixed with 1% coconut oil was used as a bait material to attract the rodents. The traps were planted over a tilted hill and on either side little quantity of the bait material was placed in order to entice the rodents to enter the trap. The hills around the trap were allowed to stand. This technique is said to improve the trapping efficiency (Neelanarayanan & Kanakasabai, 2000). The trapper(s) took nearly one to three hours to place all the traps inside the rice fields. The cost for trapping one rodent was then calculated to determine whether this technique is economical or not for the marginal and small-scale farmers and also to verify its applicability in future all over the country.

RESULTS & DISCUSSION

The number of rodents trapped using *Tanjore* bow-trap during *Kuruvai* season paddy crop month-wise and developmental stage-wise are provided in Tables 1 and 2, respectively. During *Kuruvai* season, the total number of *Tanjore* bow-traps set was 6,440. All the rodent species, that inhabit the rice fields *viz., Bandicota bengalensis, Millardia meltada* and *Mus booduga* were trapped and killed by this method (Table 1). Of the three species, *B. bengalensis* was the predominant rodent pest of rice crop of our area and 838 of them were trapped. The number of other two species of rodents *viz., M. meltada* and *M. booduga* trapped was 118 and 6, respectively (Tables 1 and 2). The overall mean trapping success in terms of percentage was 11.87, 15.17 and 17.69 during July, August and September 2006, respectively (Table 1). The mean percentage of trapping success was found to be higher during maturation stage (17.94) than milky (14.35) and vegetative stage (11.1) (Table 2).

Table 1. Number of rodents trapped using *Tanjore* bow-trap during *Kuruvai* Season paddy crop

Month & Year	*B. bengalensis*		*M. meltada*		*M. booduga*		Total number of traps set	Overall mean % of trapping success
	No. of murids trapped	Mean % of trapping success	No. of murids trapped	Mean % of trapping success	No. of murids trapped	Mean% of trapping success		
July 2006	120	10.75	26	2.16	0	0	1200	11.87
August 2006	356	12.11	80	2.82	3	1.5	3040	15.17
September 2006	362	21.58	12	1.24	3	0.41	2200	17.6
Total	838	14.81	118	6	0.63	6.440	15.2	44.64

Table 2. Number of rodents trapped using *Tanjore* bow-trap during different developmental stages of *Kuruvai* season paddy crop.

Developmental stages	*B. bengalensis*		*M. meltada*		*M. booduga*		Total.no of traps set	Overall Mean % of trapping success
	N.M.T[1]	Mean[2]	N.M.T[1]	Mean[2]	N.M.T[1]	Mean[2]		
Vegetative	170	11.25	22	2.1	-	-	1350	11.1
Milky	375	14.6	92	2.7	4	1.7	3200	14.35
Panicle formation	293	19.6	4	1.8	2	0.3	1890	17.94
Total	838	15.1	118	2.2	6	1	6440	14.46

[1] Number of murids trapped

[2] Mean percentage of trapping success

The trapped numbers of rodents using *Tanjore* bow-trap during *Samba* season paddy crop month-wise and developmental stage-wise are provided in Tables 3 and 4, respectively. Total number of traps set during this season was 14,746. Of the three inhabiting species, *B. bengalensis* was trapped to the tune of 2273 followed by *M. meltada* (36 nos.) and *M. booduga* (6 nos.). During *Samba* season largest number of traps was set during vegetative phase (5646) of the crop. However, mean percentage of trapping success was found to be higher during milky stage (17.18) followed by maturation (15.98) and vegetative (13.6) phases. Of the three months of trapping trials, January 2007 recorded maximum mean number of trapping success. Some local tribals are reported to eat trapped *B. bengalensis* and *M. meltada* and for them this is an important, high protein food supplement.

It is evident from the results that an increasing trend *i.e.,* more numbers of rodent trappings could be observed during successive developmental stages of the crop (from vegetative to maturation stage) and it might be due to the appearance and maturation of grains. Further, it is obvious from the results that this trap is to be used during the first three developmental stages of both seasons of the paddy crop in order to avoid major depredation due to rodent pests (Tables 1-4).

Table 3. Number of rodents trapped using *Tanjore* bow-trap during *Samba* season paddy crop.

Developmental stages	*B. bengalensis*		*M. meltada*		*M. booduga*		Total no. of traps set	Overall Mean% of Trapping success
	N.M.T[1]	Mean[2]	N.M.T[1]	Mean[2]	N.M.T[1]	Mean[2]		
December2006	1269	14.59	22	0.93	2	0.54	8646	14.61
January 2007	907	17.2	5	0.59	1	1.25	5220	17.3
February 2007	97	11.35	9	2.08	3	0.93	880	12.70
Total	2273	14.38	36	1.2	6	0.90	14,746	14.87

[1] Number of murids trapped
[2] Mean percentage of trapping success

The results of the *Tanjore* bow-trap-trapping trials in *Kuruvai* and *Samba* season paddy crop fields are presented in Table 5. During *Kuruvai* season 26 ha rice fields were subjected to the trials for 66 days and during *Samba* season 54 ha rice fields were covered in 79 days. The total numbers of trap nights during Kuruvai and Samba seasons' rice crop were 6,440 (range/day-80 to 240) and 14,746 (range/day -120 to 720), respectively. Three species of field rodent pests *viz., Bandicota bengalensis, Millardia meltada,* and *Mus booduga* were trapped during both seasons of rice crop. During both seasons *B. bengalensis* were trapped more in number followed by *M. meltada* and *M. booduga*. The percentage of trapping success was found to be 15 (*Kuruvai* season) and 16 (*Samba* season). The total amount spent by the farmers was Rs. 5162 for *Kuruvai* season and Rs. 12,115 for *Samba* season. The cost of trapping a single rodent was computed to be Rs. 5 during both seasons. From the results of the present study, it can be suggested that trapping can be an effective, eco-friendly method for controlling rice field rodents. Further, it should also be borne in mind that when the rodent density is high in the crop fields, the trapping efficiency would increase further, and that will significantly reduce the trapping cost per rodent. The results of the present study corroborate the findings of Neelanarayanan & Kankasabai (2000).

Table 4. Number of murids trapped using *Tanjore* bow-trap during different developmental stages of *Samba* season paddy crop.

Developmental stages	*B. bengalensis*		*M. meltada*		*M. booduga*		Total.No of	Overall Mean % of
	N.M.T[1]	Mean[2]	N.M.T	*Mean[2]	N.M.T[1]	Mean[2]	traps Set	Trapping success
Vegetative	759	13.31	20	0.94	1	0.25	5646	13.6
Milky	852	17.0	5	0.69	2	1.04	5180	17.18
Panicle formation	662	15.61	11	1.45	3	0.93	3920	15.98
Total	2273	15.30	36	1.02	6	0.74	14,746	15.58

[1] Number of murids trapped
[2] Mean percentage of trapping success

Therefore, it is deemed that this tool is a simple, economical and effective one to combat rodent menace in the crop fields. Further, with this method the traditional trappers of this area could be able to trap and kill as many as 3,277 rodents in two seasons of rice crop and earn Rs.17,727/-.

Table 5. Results of the trials of *Tanjore* bow-trap in *Kuruvai* and *Samba* season paddy

	Paddy Crop Seasons	
	Kuruvai	*Samba*
Months of Investigation	July – September 2006	December – February 2006- 07
Total Area Covered	26 ha	54 ha
Total No. of Trapping days	66	79
Total number of Traps set	6,440 (80 – 240) [1]	14,746 (120 – 720) [1]
Total no. of murids trapped	962 (0 – 43) [2]	2315 (5 – 117) [2]
No. of murids trapped /ha	37	43
Overall % of Trapping success	14.93	15.69
Rodent Species Composition (in numbers)		
B. bengalensis	804 (83.5%) [3]	2773 (98.18%) [3]
M. meltada	152 (15.8%) [4]	36 (1.55%) [4]
M. booduga	6 (0.62%) [5]	7 (0.30%) [5]
Total amount (in Rs.) earned by the trappers	Rs. 5612/-[6]	12,115/-[6]
Cost of each rodent trapped	Rs.5/-	Rs.5/-

1 Range values of total number of traps set/day

2 Range values of total number of animals trapped/day

3,4, 5 values within the parenthesis indicate the percentage of trapped rodent species composition

6 Earning has been calculated as follows: Rs.200/- for setting of 300 traps/day Rs. 1/- for every trapped rodent.

This method may very well be encouraged among peasant community. There is a large demand for the dried rodent-meat for livestock, poultry and fish for feeding. Besides, the rodent's skin has also proved to be good for making small leather products. Therefore, the trapped rodents may be used for these purposes.

It is suggested that the traditional *Tanjore* bow-trappers be identified by the Central and State agricultural personnel and Non-Governmental Organizations and be encouraged to practice this eco-friendly method. We sincerely hope that this kind of action will certainly help the marginal and small-scale farmers to a great extent to keep the rodent pests under check in their rice fields.

Advantages of this method: It requires less man power; it is a simple, economical and effective method; it does not affect the non-target animals, it is eco-friendly, it offers good trappability, all the ravaging rodent species are trapped and killed, the trapped rodents form a good source of protein for human beings, livestock etc. The number of dead rodents is actually seen by the farmers, and it brings prosperity to the traditional trappers (Neelanarayanan & Kankasabai, 2000).

ACKNOWLEDGEMENTS

The authors are thankful to Dr. S. Ramalingam, Principal and Management of Nehru Memorial College (Autonomous) for having rendered necessary facilities. We are also thankful to the rodent trappers who helped us to trap the rodents for these experiments. The moral support and encouragement extended by Prof. K. Kanagasundari, H.O.D. of Zoology, Nehru Memorial College (Autonomous), Puthanampatti are gratefully acknowledged.

References

Hegdal, P. L. & R. W. Blaskiewiez. 1984. Evaluation of the potential hazard to barn owls of Talon (Broadifacoum bait) used to control rats and house mice. *Environ. Toxicol. Chem.*3: 167-179.

Hegdal, P. L. & B. A. Colvin. 1988. Potential hazard to eastern screech-owls and other raptors of Broadifacoum bait used for vole control in orchards. *Environ. Toxicol. Chem.* 7: 245-260.

Malhi, C. S. & Chander Sheikher. 1989. Some social and ecological considerations for rodent control in District Pauri of Garhwal Himalayas. *Tropic. Pest Manage.* 35(1):67-69.

Neelanarayanan, P. 1997. *Predatory pressure of Barn owl (Tyto alba stertens* Hartert, 1929) *on Rodent pests - A Field Evaluation.* Ph.D. Thesis, Bharathidasan University, Tiruchirappalli.

Neelanarayanan, P. & R. Kanakasabai. 1995. Effect of zinc phosphide on Intestine, Liver, Spleen and Kidney of *Millardia meltada* (Gray). *J. Ecobiol.* 7(2): 85-90.

IBID. 2000. Rodent Pest Management - A Traditional Approach. *Pestology* 24 (2): 12-19.

Neelanarayanan, P., R. Nagarajan & R. Kanakasabai. 1994. Possibility of secondary poisoning hazard to house crow, *Corvus splendens* and jungle crow, *Corvus macrorhynchos. Zoo's print* 9(5): 16.

IBID. 1996. Burrow morphology of field rodents. *J. Bombay Nat. Hist. Soc.* 93(2): 238-241.

53

Efficacy of Herbal Products Against Pigeon Pea Pod Borers

NANDIHALLI, B. S.

ABSTRACT

The experiment on efficacy of three herbal products against pigeon pea pod borer carried out along with profenophos at Main Agricultural Research Station, Dharwad during 2005-06 indicated that among three herbal products Cristol 56 SL III was found effective in reducing the pod borer population with less pod borer damage and higher grain yield which was camparable with profenophos.

INTRODUCTION

Pigeon pea is an important pulse crop widely grown in the country. Among the pigeon pea growing states, Karnataka ranks fourth in area (4.4 lakh ha) and fifth in production (2.2 lakh tones) with a productivity of 541 kg/ha. Among many factors responsible for low yields of pigeon pea, insect pests are the major ones. Of these, the pod borers (*Helicoverpa armigera, Maruca testulalis,* etc.) are the most serious endemic pests. The pesticide usage on pigeon pea has been both extensive and intensive (Gour, 1993). This is evident from the fact that 4% of pesticide consumption is by pulses and out of which 80% is consumed by pigeon pea. This has lead to the problem of insecticide resistance resulting in failure of insecticide control programme.

The various natural products such as neem, pongamia, *Vitex negundo,* garlic etc. possess an array of unique phytochemicals having diverse biological effects on the pests (Jayraj & Ingacimuthu, 2005). These herbal products are ecofriendly and constitute one of the important IPM tools. In view of this, the herbal products Cristol (extracted from Garlic and Pongemia oil; the earlier name of the product was KK product), developed from Krishna Antioxidants Pvt. Ltd., Mumbai were evaluated in the management of pigeon pea pod borers.

MATERIAL & METHODS

The experiment on efficacy of herbal products against pigeon pea pod borer was carried out in randomized block design at Main Agricultural Research Station, Dharwad during 2005-06. Three herbal products namely Cristol 56 SL I, II and III along with standard chemical and untreated control with four replications were evaluated against pod borers using Maruti variety of Pigeon pea. The crop was sown in a plot size of 3.6 X4.5 m for each treatment in the month of July 2005 with spacing of 90 X 20 cm between rows and plants, respectively. All agronomic practices were followed except plant protection measures in raising the crop.

Observations were recorded on number of pod borer larvae per plant on randomly selected 10 plants in each treatment on a day before, 2, 7 and 15 days after first and second sprayings. At the time of harvest, number of healthy pods, damaged pods per plant and grain yields was recorded per plot and yields were converted on hectare basis. The data were subjected to statistical analysis.

RESULTS & DISCUSSION

There was no significant difference in recorded number of Red gram pod borer larvae/plant at 1 DBS during first spray. At 2 DAS, Cristol 56 SL (III) (4.07 larvae/plant) recorded significantly less number of pod borer larvae per plant which was on par with Cristol 56 SL (II) and profenophos 50 EC. Among Cristol 56 SL products, Cristol 56 SL (II) recorded less number of larvae at 7 DAS, which was on par with profenophos. At 15 DAS all three herbal products were on par with each other (Table 1).

Table 1. Evaluation of CRISTOL 56 SL against red gram pod borers during first Spray (2005-06)

Treatment	Dosage (ml/l)	No. of larvae per plant			
		1DBS	2 DAS	7DAS	15 DAS
1. CRISTOL 56 SL (I)	5.00	5.00	6.93	3.07	0.60
2. CRISTOL 56 SL (II)	5.00	3.80	5.10	1.20	0.47
3. CRISTOL 56 SL (III)	5.00	6.07	4.07	3.53	0.33
4. Profenophos	2.00	3.60	5.23	1.17	0.33
5. Untreated control	-	4.87	7.00	4.17	1.33
CV %		15.07	15.06	13.51	11.26
CD		NS	2.33	2.51	0.32

During second spray, Cristol 56 SL (III) recorded less number of larvae per plant (1.60 larvae/ plant) at 2 DAS. All three Cristol 56 SL products were on par with each other at 7 DAS. However, profenophos recorded significantly less number of larvae per plant (1.60 larvae / plant) at 7 DAS. All three Cristol 56 SL products were on par with profenophos at 15 DAS (Table 2).

Table 2. Evaluation of CRISTOL 56 SL against red gram pod borers during second spray (2005-06) DBS = Day before spray DAS = Day after spray

Treatment	Dosage (ml or g/l)	No. of larvae per plant			
		1DBS	2 DAS	7DAS	15 DAS
1. CRISTOL 56 SL (I)	5	4.33	2.93	2.33	0.28
2. CRISTOL 56 SL (II)	5	3.60	3.27	2.40	0.26
3. CRISTOL 56 SL (III)	5	4.20	1.60	2.20	0.23
4. Profenophos	2.0	4.20	2.73	1.60	0.40
5. Untreated control	-	4.00	4.13	3.13	1.27
CV %		17.88	14.68	14.37	11.41
CD		NS	1.22	0.60	0.32

Among three Cristol 56 SL products, Cristol 56 SL (III) recorded significantly higher number of good pods per plant (39.30/plant) which was on par with profenophos. All three Cristol 56 SL products were on par with each other in recording less number of damaged pods. However, profenophos recorded significantly less number of damaged pods per plant (10.50 damaged pods/plant). Similar trend was observed with respect to percentage of damaged pods. Among three Cristol 56 SL products, Cristol 56 SL (III) recorded highest yield/ ha (13.53 q/ha) which was on par with profenophos and other two Cristol 56 SL (Table 3). The present findings are in agreement with the report of Singh *et al.*, (2004) who observed the highest efficacy of botanicals with insect mortality up to 80.7%.

Table 3. Effect of CRISTOL 56 SL on Red gram pod borer damage and yield (2005-06)

Treatment	Dosage (ml/l)	No. of pods /plant		% of pods damaged	Yield in q/ha
		Good	Damaged		
1. CRISTOL 56 SL (I)	5.00	33.1	15.3	31.72	12.24
2. CRISTOL 56 SL (II)	5.00	34.9	14.4	28.90	13.01
3. CRISTOL 56 SL (III)	5.00	39.3	14.3	26.86	13.53
4. Profenophos	2.00	40.5	10.5	20.31	13.89
5. Untreated control	-	23.9	28.8	54.65	11.22
CV %		13.85	15.83	12.82	7.00
CD		8.20	6.56	7.59	1.52

In conclusion, Cristol 56 SL III was found promising and comparable with profenophos in the management of red gram pod borer. Hence, the herbal product Cristol can serve as an alternate to the chemicals in the management of pod borer.

References

Gour T.B. 1993. Use and abuse of insecticides with special reference to synthetic pyrethroids on pulses and oilseeds. In: *Compendium of lectures of summer institute on advances in entomology with special reference to oilseeds and pulses.* 17th May to 5th June, 1993 held at Andhra Pradesh Agricultural University, Hyderabad. pp. 325-341.

Singh, B., Sunil Kumar & R. K. Akhauri. 2004. Efficacy of some biorational insecticides on larval population of gram pod borer on pigeon pea. *Legume Research* 27: 62-65.

54

Management of Thrip Complex in Grape Ecosystem

SUNITHA, N. D., S. B. JAGGINAVAR
D. R. PATIL AND A.P .BIRADAR

ABSTRACT

Studies were conducted on the management of thrips in grape with nine treatments comprised of three neo-necotinoids namely acetamaprid 20% SP, Imidacloprid 17.8% SL and thiamethoxam 25%WG, acephate 75%SP, oxydemeton methyl25%EC (standard check) azadirachtin 0.03%, garlic chilli extract in kerosene, water spray and control in randomized block design replicated 4 times with six vines per treatment at Regional Horticultural Research Station, Bijapur, Karnataka. All the three necotinoid molecules were found significantly superior over standard check in percent reduction of thrip population (94.41, 96.04, 95.76) which reflected in reduced percent bunch infestation (17.20, 16.69 & 14.90), yield per acre (101.36, 90.38 & 88.86 qt) and C:B ratio (1:3.76, 1:3.34 & 1:3.11 respectively) as against standard check which recorded 84.06 per cent pest infestation, 41.45 per cent bunch infestation, 63.51 quintals of berry yield and C:B ratio of 1:2.09 . Acephate 75% Sp was found on par with neo necotinoids and azadirachtin 0.03 per cent was found on par with standard check. Garlic chilly extract in kerosene and water spray were found significantly inferior to standard check.

INTRODUCTION

Grape is one of the important fruit crops of India used for table purpose, resin and wine making with good medicinal value due to presence of large amount of antioxidants. In India, the current area and production under grape is estimated at 57,800 ha with an annual production of 147,2800 tones. Maharastra accounts for more than 80 per cent of the production and Karnataka stands second after Maharastra (Chadha, 2006). In Karnataka, grape is cultivated in an area of 8500 ha with an annual production of 1.20 lakh tones. Bijapur is one of the major districts of Karnataka in grape cultivation (5000 ha) with an annual production of 1 lakh tones. Along with diseases, insect pests are the important production constraints in grape cultivation. Large number of insect pests attack grapevine throughout the year since it is perennial.

Nearly 85-94 species of insect pests of grape have been reported in India (Butani, 1979; Tandon & Verghese 1994; Atwal & Dhaliwal, 2005). Among them thrips are the most destructive sucking insects of grapevine which damage foliage, blossoms as well as berries.

The attacked foliage from the species like *Rhipiphorothrips cruentatu* Hood take whitish hue, acquire a withered appearance and then turn brown. Leaves ultimately curl up and drop off. Such vines either do not bear fruits or fruits drop off prematurely. The matured fruits are of poor quality. The berry thrips like *Scirtothrips dorsalis* (Hood) lacerate and suck the sap from the berries of all stages leading to the formation of black corky surface resulting in scab formation (Reddy, 1957) which affects marketability and export as visual appeal is the basic requirement in marketing table grapes. *Scirtothrips dorsalis* is the most important pest causing damage to berries and 10th to 12th leaves host more thrip population compared to other leaves (Rangareddy & Murthy, 2006). Three species of thrips namely, *Thrips hawaensis*, *Thrips palmi* and *Scirtothrips dorsalis* were found infesting grape vine in and around Bijapur district. Looking in to the severity of the problems, studies on management of thrips in grape were taken at Regional Horticultural Research Station, Bijapur during 2006-07.

MATERIAL & METHODS

Studies were conducted during 2006-07 with nine treatments (Table 1) replicated four times with six vines for each treatment. The variety under study was Thompson seedless. The first spray was imposed during first week of June exactly one month after April-May pruning. Second spray was taken one month after the spray with a spray solution of 1000 liters per hectare. Pre-counts of thrips per three fresh leaves were taken one day before spraying. Subsequent observations were taken 3, 7 and 10 days after the imposition of treatments. The average per cent reduction in thrips population at different intervals was worked out using the Henderson & Tilton (1955) formula as given below.

$$\text{Percent reduction} = 100 \left\{ 1 - \frac{Ta \times Cb}{Tb \times Ca} \right\}$$

Where Ta=number of insects in the treated plot after insecticide application, Tb= number of insects in the treated plot before insecticide application, Cb=Number of insects in the untreated check before insecticide application and Ca= Number of insects in the untreated check after insecticide application.

The percent reduction was transformed to angular values, from which analysis of variance was calculated for determining critical difference (C.D) at 5 percent level of significance.

Studies were continued between October prunings to harvesting wherein the first spray was given at berry setting stage. Subsequent spray was given one month after first spray. Like previous trial, pre-counts of thrips were taken. Observation on number of infested bunches were recorded. Bunch with more than ten berries showing scab symptoms were considered as infested bunch. Later per cent infestation was calculated. The data was subjected for statistical analysis after suitable transformation. Yield data is recorded as kg/vine and yield per acre was computed and the C:B ratio was calculated by taking into consideration the annual management cost of the orchard, insecticide cost for four sprays, and price of marketable fruit.

RESULTS & DISCUSSION

It is evident from table 1 that all the three neo necotinoids *viz,* acetamaprid 20% SP, imidacloprid 17.8% SL and thiamethoxam 25% WG were significantly superior over standard check oxydemeton methyl 25% EC and on par with each other in reducing the thrip population at different intervals after application. Percent reduction showed increased trend from 3 days towards 10 days after spraying and at 10 days they recorded 94.41, 96.04 and 95.76% reduction in pest population, 17.20, 16.69, and 14.90% bunch infestation respectively. This was reflected in yield of berries (101.36, 90.38 and 88.56 QHs/acre) and cost benefit ratio (1:3.76, 1:3.34 and 1:3.11 respectively) (Table-2). Acephate 75% Sp performed on par with the neonecotinoids which recorded 93.38% reduction in thrip population, 76.68 qtls of berry yield and C:B ratio of 1:2.73. Standard check oxydemeton methyl 25% EC and azadirachtin 0.03% were found on par with each other and recorded 84.06 and 86.45 per cent reduction of the pest, 63.51 and 54.76 quintals of fruit yield and C:B ratio 1:2.09 and 1:1.59 respectively. Garlic chilly extract and water spray were found significantly inferior to all other treatments (Table-2).

The present findings of effectiveness of neo necotinoids on the management of thrips are in agreement with the findings of Kulkarni & Adsule (2006a & b) who obtained good results of imidacloprid 200% SL @ 3ml/10 litre and thiamethoxam 25%WG@ 25gm a.i/ha against grape thrips. Apart from this several workers have reported effectiveness of these chemicals on thrips and other sucking insects in various ecosystems. Similar results on the effectiveness of imidacloprid in reducing sucking pests were observed by Verma *et al.,* (2003) and Sreekanth *et al.,*(2004) against jassids and thrips, Burleigh *et al.,* (1998) and Seal *et al.,* (1994) against chilli thrips, Jain & Ameta (2006) against chilli thrips and whiteflies. In the same manner high susceptibility was reported for many sucking insects like leafhoppers, *Amrasca biguttula biguttula, A. divastans* (Patil *et al.,* 2004; Mishra, 2002; Mathirajan & Regupathy, 2001), *Neelaparvatha lugens* and *Sogetella furcifera* (Mathirajan & Regupathy, 2002), *Thrips tabaci* (Praveen, 2003), Sugarcane woolly aphid (Vijayaraghavan & Regupathy, 2006), to thiamethoxam and imidacloprid. Sinde *et al.,* (2003) reported the effectiveness of acetamaprid, thiamethoxam and imidacloprid against sugarcane woolly aphid.

From the present investigations it can be concluded that acephate and the three neo-necotinoids can be used in the management of thrips in grape ecosystem considering the pre-harvest waiting periods set for each insecticide (60 days for imidacloprid, forty days for thiamethoxam, three days for azadirachtin etc.) and stage of the crop. Insecticides with minimum waiting period are recommended as in the case of infestation between October prunings to harvesting of berries.

Table 1. Reduction in thrip population as influenced by different treatments

S, No	Treatments	Dose/lit	Thrips / 3leaves 1 DBS (Pooled)	Per cent reduction								
				3DAS			7DAS			10DAS		
				1st spray	2nd spray	Pooled	1st spray	2nd spray	Pooled	1st spray	2nd spray	Pooled
1	Acetamaprid 20% SP	0.3gm	18.40	57.53 (49.41)	65.64 (54.09)	61.58 (51.71)	80.39 (63.72)	80.82 (64.04)	80.49 (63.65)	95.30 (77.53)	93.53 (75.31)	94.41 (76.31)
2	Imidacloprid 17.8% SL	0.3ml	18.79	51.24 (45.71)	69.96 (56.82)	60.60 (51.12)	82.19 (63.09)	82.70 (65.46)	82.44 (65.27)	95.52 (77.94)	96.57 (79.37)	96.04 (78.65)
3	Thiamethoxam 25% WG	0.3gm	17.54	56.28 (48.62)	63.95 (53.16)	60.11 (50.83)	86.17 (68.19)	79.74 (63.30)	82.95 (65.57)	94.16 (76.21)	97.37 (80.68)	95.76 (78.03)
4	Acephate 75% SP	1.0gm	17.23	55.92 (48.39)	54.30 (47.47)	55.11 (47.94)	80.02 (63.54)	84.36 (66.73)	82.19 (65.05)	93.72 (75,48)	94.04 (76.03)	93.88 (75.70)
5	Oxydemeton methyl 25% EC (Standard check)	1.0ml	15.30	51.89 (46.08)	49.86 (44.91)	50.87 (45.50)	69.32 (56.38)	82.52 (65.37)	75.92 (60.87)	79.18 (62.87)	88.94 (70.72)	84.06 (66.50)
6	Azadirachtin 0.03 %	5.0ml	16.12	44.36 (41.74)	48.33 (44.04)	46.34 (42.89)	65.58 (54.08)	78.25 (62.17)	71.91 (58.20)	86.18 (66.86)	86.72 (68.61)	86.45 (68.36)
7	Garlic chili kerosene extract	5.0ml	15.33	39.18 (38.67)	50.68 (45.39)	44.93 (42.03)	63.20 (52.71)	72.60 (58.44)	67.90 (55.49)	73.50 (59.02)	51.66 (45.95)	62.58 (52.30)
8	Water spray	-	16.70	14.30 (22.22)	06.24 (14.42)	10.27 (18.63)	08.25 (15.98)	03.50 (8.52)	05.87 (14.81)	8.25 (16.95)	3.89 (11.09)	6.07 (14.30)
9	Untreated check		16.62	0.10 (1.81)	0.10 (1.81)	0.10 (1.81)	0.10 (1.81)	0.10 (1.81)	0.10 (1.81)	0.10 (1.81)	0.10 (1.81)	0.10 (1.81)
	SEm±		1.318	1.42	2.36	1.81	1.97	1.89	1.81	4.74	1.96	3.13
	CD at 5%		NS	4.26	7.08	5.16	5.91	5.65	5.18	14.20	5.87	8.93
	CV%		14.14	6.30	10.00	8.86	6.91	6.35	8.86	13.35	5.99	12.98

DAS: days after spraying DBS: Days before spraying Figures in parenthesis are arc sine transformed values

Table 2. Per cent bunch infestation, yield and economics of thrips management in grape

S. No	Treatments	Dose/ l	Thrips/ 3leaves (Pooled)	Total bunches/ Vine	Percent bunch Infestation	Yield / vine (Kg)	Yield/ acre (Qt)	Gross returns	Cost of thrips management Per acre (Rs)	Total cost (Rs)	Net Returns (Rs)	C:B ratio
1	Acetamaprid 20% SP	0.3gm	15.83	27.00	17.20 (24.50)	9.33	101.36	152,040	1,920	31920	120120	1:3.76
2	Imidaclоprid 17.8% SL	0.3ml	16.08	28.41	16.69 (24.12)	8.30	90.38	135570	1,200	31200	104370	1:3.34
3	Thiamethoxam 25% WG	o.3gm	15.75	26.66	14.90 (22.26)	8.16	88.86	133290	2,400	32400	100890	1:3.11
4	Acephate 75% SP	1.0gm	15.08	27.00	27.68 (31.52)	7.06	76.88	115320	896	30896	84424	1:2.73
5	Oxydemeton methyl 25% EC	1.0ml	13.75	26.00	41.45 (40.02)	5.86	63.51	95265	768	30768	64497	1:2.09
6	Azadirachtin 0.03 %	5.0ml	12.58	25.00	48.59 (44.18)	5.03	54.76	82140	1,600	31600	50540	1:1.59
7	Garlic chili kerosene extract	5.0ml	12.00	26.00	61.00 (51.35)	2.26	24.61	36915	640	30640	6275	-
8	Water spray		13.08	24.08	69.92 (56.78)	1.36	14.81	22215	-	30000	-	-
9	Control		12.33	28.58	84.93 (67.39)	1.23	13.39	20085	-	30000	-	-
	SEm±		1.47		2.02	0.359	3.898					
	CD at5%		NS	NS	6.06	1.075	16.00					
	CV%		18.11		8.89	11.50	11.53					

References

Atwal, A. S. & G. S. Dhaliwal. 2005. *Agricultural Pests of South Asia and their management*. Kalyani Publishers, New Delhi. Pp.307-308.

Burleigh, J. R, V. Kusingham, W. R. B. Lalith & S. Gonapinuwala. 1998. Pattern of pesticide use and efficacy among chilli growers in dry zone of N.E.Sri Lanka (System B): perception v/s reality. *Agriculture, Ecosystem and Environment*.701:49-60.

Butani, D. K. 1979. *Insects and fruits*. Periodical Expert Book Agencies, New Delhi. 160pp.

Chadha, K. L. 2006. Indian Viticulture Scenario. *In: International symposium on grape production and processing*. Feb. 6-11, 2006, Baramati, India. Pp. 12-13.

Henderson, C. F. & E. W. Tilton. 1955. Tests with the acaricides against brown wheat mite. *J. Econ. Entom.* 48 (2): 257-261.

Jain, K. L. & O. P. Ameta. 2006. Bio-efficacy of imidacloprid and Beta-cyflothrin against insect pests of chilli. *Pestology* 30 (1): 36-39.

Kulkarni, N. S. & P. G. Adsule. 2006a. Effect of confidor 200 SL (Imidacloprid) for the management of flea beetle, thrips and jassids of grape. *Pestology* 30 (4): 52-56.

IBID 2006b. Efficacy of Actara 25 WG (Thiamethoxam) on the management of flea beetle, jassids and thrips in grape. *Pestology* 30 (5): 13-17.

Mathirajan, V. G. & A. Regupathy. 2001. Persistent toxicity of thiamethoxam to *Apis gossypii* Glover and *Amrasca devastans* Distant on cotton. *Pest Management and Economic Zoology* 9 (2): 155-160.

IBID, 2002. Toxicity of thiamethoxam to *Apis gossypii, Amrasca devastans, Neelaparvatha lugens and Sogetella furcifera. Annals of Plant Protection Science* 10 (2): 369-370.

Mishra, H. P. 2002. Field evaluation of some newer insecticides against aphids (*Aphis gosypii)* and jassid (*Amrasca bigutttula biguttula*) on okra. *Indian J.Entomol.* 64 (1):82-84.

Patil, S. B., S. S. Udikeri & B. M. Khadi. 2004. Thiamethoxam 35 FS-A new seed dresser formulation for sucking pest control in cotton crop. *Pestology* 28 (3):34-37.

Praveen, P. M. 2003. *Studies on insecticide resistance in early season sucking pests of cotton in Tamil Nadu*. Ph. D. Thesis. Tamil Nadu Agricultural University, Coimbatore.

Rangareddy, A. & P. R. Murthy. 2006. Incidence levels of berry thrips, *Scirtothrips dorsalis* Hood as influenced by site of infestation and identification of resistance sources through varietal screening. *International symposium on grape production and processing*, Feb. 6-11, 2006, Baramati, India.127pp.

Reddy, D. B. 1957. Scab on grapes. *Proceedings of 44th Indian Science Congress*-III. Pp. 393.

Seal, D. R., R. M. Barenowski. & J. D.Bishop. 1994. Effectiveness of insecticides in controlling *Thrips palmi* Karny. *Proceedings of Florida state Horticultural Society*. 160: 228-233.

Sinde, V. D., S. B. Magar, S. S. Patil & M. B. Deshmukh. 2003. Evalaution of insecticides for the control of white woolly sugarcane aphid, *Ceratovacuna lanigera* Zchht. *In 52nd Annual Convention of the Deccan Sugar technologies association (India)*, Pune, India.

Sreekanth, M, M., R.D. Sreeramalu, B.S. Rao, & T. R. Babu. 2004. Evalaution of certain new Insecticides against *Thrips palmi* Karny, the vector of peanut bud necrosis virus (PBNV) on mung bean (*Vigna radiata* LWilczek). *International Pest Control* 46 (6): 315-317.

Tandon, P. L. & A. Verghese. 1994. Present status of insect and mite pests of grapes in India. *Drakshavritta* 149-157.

Verma N. R. G., S. M. Zaheruddeen, B. Bhavani & P. R. M. Rao. 2003. Efficacy of certain new insecticides against rice plant hoppers under field conditions. *Indian J. plant prot*. 31(2):31-33.

Vijayaraghavan, C. & A. Regupathy. 2006, Bio-efficacy of Thiamethoxam (Actara 25 WG) against sugarcane woolly aphid. *Pestology* 30 (1):31-35.

55

Testing of Bt Products for the Management of Greater Wax Moth, *Galleria mellonella* L

VISHWAS, A.B. AND GAVIGOWDA

ABSTRACT

The Greater wax moth, *Galleria mellonella* is considered as a notorious pest of honeybee colonies and is well distributed throughout the world. The management of this pest is a difficult task because no chemicals can be used on this pest as the honey bees are involved. Therefore biocontrol method is most ideal for management of greater wax moth. Since its discovery in 1915 in Germany, the bacterium, *Bacillus thuringiensis Bt* has been tried in insect control and was found to cause mortality in the larvae of Diptera, Coleoptera and Lepidoptera. With this view testing of *Bt* products was carried out to control *G. mallonellla*. One *Bt* culture (UAS strain) which was isolated and found effective on greater wax moth was compared with three commercial formulations namely Halt, Dipel 8L and Biolep for their efficacy against different instar larvae of the greater wax moth at the dosages of 500, 1000, 2500 and 5000 ppm per 100 g of artificial diet under lab conditions. All the formulations were found toxic to wax moth larvae and differed significantly in their efficacy. The extent of mortality depended on the age of the larvae causing up to 100 percent mortality in 1st instar larvae and 30.0 to 90.0 percent death in the 7th instar larvae. The comparative efficacy of four formulations of *B. thuringiensis* against third instar larvae of *G. mellonella* was studied in the laboratory. On the basis of LC_{50} it was observed that UAS strain was most toxic (LC_{50} 805.04 ppm) followed by Dipel 8L (LC_{50} 1071.2 ppm), Halt (LC_{50} 1029.3 ppm) and Biolep (LC_{50} 1251.7 ppm). The LT_{50} values for third instar larvae revealed that UAS strain was able to kill the larvae at a faster rate recording LT_{50} of 2.6 days at 2500 ppm, followed by 3.84, 3.90 and 4.02 days for Halt, Dipel and Biolep respectively. All the *Bt* formulations tested were found to be highly toxic to silkworm causing 100 percent mortality of fourth instar larvae within five days and the same *Bt* formulations when fed to adult worker bees through 50 percent honey solution in cages were found safer to honey bees. No adverse effect was observed. The empty combs treated with *Bt* were well protected from wax moth infestation during one month period as compared to untreated combs.

56

Bat Species Survey-an Effective Tool for Bat Conservation in Agasthiyamalai Biosphere Reserve

JULIET VANITHARANI AND U.S.U. MALATHI

ABSTRACT

Western Ghats which extend through the southern states of peninsular India are a treasure house of species diversity. Parts of southern Western Ghats, the Agasthiyamalai biosphere reserve are Internationally recognized for very rich biodiversity. Being a hotspot they provide habitat for endemic flora and fauna. Our bat survey has revealed occurrence of 33 bat species out of 8 families which includes 6 Megachiropterans and 27 Microchirapterans. The available bat diversity in the Agasthiyamalai biosphere reserve is an important component to maintain the eco-balance because of the vital role it plays namely of controlling insect pests, dispersing seeds and pollinating flowers. Disturbance of these highly efficient bio-agents will adversely affect the whole ecosystem. Conservation of this mammalian species is one of the important aspects for forest restoration and the maintenance of agro-ecosystem. This paper is an attempt to recommend these poorly recognized bat species for protection and conservation through Indian Government legislations.

57

Relationship Between Frequency of Bee Visits and Seed Set in Niger

BHAT, N. S., C.J. NAGARAJA, S. SUMA AND A.B. VISHWAS

ABSTRACT

The role of honey bees in the seed set of Niger, an important minor oil seed crop of India was investigated. The visits of two species of honey bees namely, *Apis cerana* and *A. dorsata* to Niger blooms was regulated to a frequency of one, two and three visits per day per capitulum by bagging the bloom and allowing above stated number of visits of each of the species daily. The seed set due to different frequencies of visits of the two honey bee species was compared with the seeds set in open pollinated capitula and capitula bagged throughout the bloom period. The treatment with one visit per capitulum per day of *A. cerana* recorded 39.2 seeds/capitulum which was on par with the seed yield of capitula subjected to open pollination (43.2 seeds/capitulum). The treatment with two visits per capitulum per day of *A. cerana* recorded 40.40 seeds and that with three visits per day recorded 41.20 seeds per capitulum which were on par with each other. The treatment with capitula closed on all days recorded significantly least number of seeds per capitulum (0.2). There was no significant difference in seeds/capitulum among treatments with one, two or three visits per day per capitulum of *A. cerana*. In case of *A. dorsata*, single visit per day per capitulum produced a mean seed yield of 37.2

seeds/capitulum which was on par with seeds set per capitulum with two (40.80) and three visits (41.4) per day per capitulum. Bagged capitula produced 0.2 seeds/capitulum. No significant difference was observed among level of seed set with differential number of visits between the two species to the Niger capitulum. That apart foraging frequency and rate of foraging by honey bee in Niger will also be elaborated.

58

Insect Biodiversity in Grape Ecosystem

SUNITHA N.D., S.B. JAGGINAVAR
D.R. PATIL AND A.P. BIRADAR

ABSTRACT

Grape being a perennial crop the vine is attacked by more than 100 insect pests at various stages of its growth in India. Studies were carried out for four years between 2003-04 to 2006-07 to understand the insect pest complex and their natural enemies at Horticulture Research Station, Bijapur and in farmer's fields in and around Bijapur. Three species of thrips namely *Thrips pami, Scirtothrips dorsalis* and *Thrips hawaensis* (Thripidae, Thysanoptera), mealy bug, *Meconellicoccus hirsutus* (Pseudococcidae, Homoptera), flea beetle, *Scelodonta strigicollis* (Eumolpidae, Coleoptera) were recorded throughout the study period in all the sample orchards as regular pests. Incidence of stem borer, *Coelosterna scabrator* (Cerambycidae, Coleoptera) was common with per cent infestation ranging between 4-65 especially in neglected gardens. Defoliator complex included adults of *S. strigicollis,* adults of root grub, *Holotrichia consanguinea,* (Scaraebidae, Coleoptera) and Sphingid caterpiller, *Acherontia styx* (Sphingidae, Lepidoptera) with less than 4% defoliation. Root feeders recorded were grubs of *S. strigicollis, H. consanguinea* and termite,

Odontotermes obesus (Termitidae, Isoptera). Honey bee, *Apis cerana indica* (Apidae, Hymenoptera), Wasp, *Vespa orientalis* (Vespidae , Hymenoptera) along with *Spodoptera litura* and *Helicoverpa armigera* (Noctuidae, Lepidoptera) were found as berry pests during ripening stage. Bostrychid, *Rhizopertha dominica* (Bostrychidae, Coleoptera) with termites was found on dead vines. The study revealed that thrips and mealy bugs as pests affect the quality production of grapes. No natural enemies were recorded except *Chrysoperla carnea* (Chrysopidae, Neuroptera).

SECTION - 6
APPLIED BEHAVIOUR

59

Spermatophore Deposition and Sperm Transfer in an Oribatid Mite *Scheloribates praeincisus interruptus* Berlese (Acari: Oribatei)

JULIE, E. AND N. RAMANI

ABSTRACT

Spermatophore deposition and sperm transfer were studied in *Scheloribates praeincisus interruptus*, a common oribatid representative well known for its ability in enhancing soil fertility. Rearing of the mite was carried out in the laboratory at 30°C and RH of 70%. Males were found to deposit spermatophores in preferred places within 3-8 days after their emergence. Spermatophore deposition was found enhanced by factors like availability of preferred food, presence of female and adequate moisture content. A single male deposited 25-40 spermattophores per day. The females initiated oviposition within 9-15 days of their emergence at concealed places on the culture substratum.

INTRODUCTION

Sperm transfer is mainly achieved in oribatid mites by the production of stalked spermatophores, which are deposited on a substratum and later are taken by the females. The deposition of spermatophores in oribatid mites was reported earlier by various authors (Sengbusch 1961; Woodring & Cook, 1962; Rockett & Woodring, 1966; Shereef, 1972; Haq & Adolph, 1981; Kummel & Dobner, 1986; Fernandez *et al.*, 1991). The present study was undertaken to acquire knowledge on the reproductive strategies of common oribatid representative viz., *Scheloribates praeincisus interruptus*. The above species was recognized as a widely distributed and common inhabitant of soil ecosystem.

MATERIAL & METHODS

Soil and litter samples underneath gooseberry trees, *Phyllanthus emblica* growing in a homeyard at Kizhisserry, Malappura District of Kerala, India were collected and brought to laboratory for extraction of mites. Collected samples were extracted in series of modified Berlese Tullgren funnels for 2-3 days depending upon the moisture content of the samples. Live mites were extracted into water/moistened litter powder while 70% alcohol was used as the extraction medium for taxonomic studies. Rearing of the mites was carried out in the laboratory in plastic containers based with a mixture of plaster of Paris and charcoal (4:1 ratios) at a temperature of 30^{0}C and RH of 70%. Different species of fungi, wood and leaf litter were offered to know the selective preference of the species to individual food items. Regular observations were made in a Stereozoom microscope under 32x magnification. Spermatophores which were deposited on the substratum, were carefully transferred on to glass slides with fine needles and passed through alcohol series-70%, 90%, 100% and stained with eosin. The stained materials were observed by using a canon digital camera attached to an Axioskop 2plus Zeiss Trinocular research microscope under 45x magnification.

RESULTS & DISCUSSION

Based on the feeding preference shown by the species, fungi like *Trichoderma harzianum* and *Curvularia geniculata* were selected as the food items for conducting developmental studies (Figs.1&2) of *S. praeincisus interruptus*. The newly emerged adult males laid large number of spermatophores on and around the food item, 3-8 days after their emergence. Woodring & Cook (1966) reported that males of *Ceratozetes cisalpinus* and *Scheloribates laevigatus* started to produce spermatophores about 10 days after emergence. Results of the present study revealed that spematophore deposition by males of *S. praeincisus interruptus* was enhanced by a combination of factors like preferred food item, presence of female and adequate moisture conditions. Such a behavioural trend was already reported in *Spatiodamaeus subverticillipes* (Shereef, 1972). A single male deposited 25-40 spermatophores per day. Males were found to deposit large number of spermatophores on the substratum, culture walls, in between faecal pellets, on the food item or on the left out agar medium. The spermatophores resembled dew drops in appearance, were stalked, bearing globular shining heads (Fig.3). Similar observations were made earlier by Sengbusch (1961) and Woodring & Cook (1962). Prior to deposition the males remained stationary and subsequently raised their body, leaving behind the spermatophore on the substratum. Several post depositional changes were noted in the spermatophore. The erect posture and glittering appearance were soon lost. Females which were moving around and over the spermatophores were found to have their genital flaps widely opened, so as to take up the globular head through their genital opening, leaving behind the stalks intact. The females initiated oviposition within 9-15 days. Concealed places on culture substratum were the preferred site for oviposition.

Fig. 1 **Fig. 2** **Fig. 3**

CONCLUSION

S. praeincisus interruptus, was found to deposit stalked spermatophores on the substratum which were later taken up by females through genital opening. The present study enabled to record the wide range of food selection of *S. praeincisus interruptus* and the results revealed higher preference of the mite for the fungi, *Trichoderma harzianum* and *Curvularia geniculata*. Further, the periods of egg maturation and oviposition were also found greatly influenced by the nutritional quality of the diet offered.

References

Fernandez, N. A., G. Alberti & G. Kummel. 1991. Spermatophores and Spermatozoa of Oribatid mites (Acari: Oribatida). Part I: Fine structure and Histochemistry. *Acarologia*. 33: 1991

Haq, M. F. & C. Adolph. 1981. Spermatophore deposition and transfer in *Pelokylla malabarica* (Acari: Oribatei). *Entomon*. 6: 135-141.

Kummel, G. & C. H. Dobner. 1986. Fine structure of spermatophores of some Oribatid mites (Acari, Arachnida). *J. Morphol.* 189: 295-311.

Rockett, C. L. & J. P. Woodring. 1986. Biological investigations on a new *Ceratozetes* and *Pergalumna* (Acarina: Cryptostigmata). *Acarologia* 29: 511-520.

Sengbusch, H. G. 1961. A study of spermatophores and sperm transfer in the Oribatei (Acarina). *J. Parasit.* 47:43-44.

Shereef, G. M. 1972. Observations on oribatid mites in the laboratory cultures. *Acarologia* 14: 281-291.

Woodring, J. P. & E. F. Cook. 1962. The biology of *Ceratozetes cisalpinus* Berlese, *Schleoribates laevigatus* Koch and *Oppia neerlandica* Oudemans (Oribatei) with a description of all stages. *Acarologia* 4: 101-137.

60

Behavioural Development of Two Forms of Shoot Bug, *Peregrinus maidis* (Ashmead) on Sorghum

VIJAY KUMAR, L, S.U.PATIL AND
A. K.CHAKRAVARTHY

ABSTRACT

Studies on various biological parameters and behavioural development of sorghum shoot bug, *Peregrinus maidis* (Ash.) revealed that, both macropterous and brachypterous forms were found in each sex. Among two forms, brachypterous ones recorded shorter incubation (7.27±0.65 days), pre-mating (6.66±1.60 hrs) and pre-oviposition (2.20±0.9 days) period as against the same being 8.35±1.05 days, 8.10±1.9 hrs and 2.40±1.0 days, respectively in macropterous form. On the other hand, brachypterous forms recorded higher mating (0.84±0.2 hr), oviposition (5.93±1.4 days) and post-oviposition (18.95±3.2 days) duration as compared to macropterous forms recording 0.61±0.2 hrs, 5.66±1.70 days and 17.87±4.07 days, correspondingly. Brachypterous forms also recorded higher fecundity (16 to 106 eggs/female) than macropterous forms (22 to 72 eggs/ female).

INTRODUCTION

Sorghum, *Sorghum bicolor* (Linn.) is the major source of food and fodder for millions of people in tropics and semi-Arid tropics of the world (Swindale, 1990). It is attacked by over 150 insect species from seedling to the harvest (Sharma, 1985). Of the different insect pests, shoot bug, *Peregrinus maidis* Ashmead (Homoptera: Delphacidae) previously considered to be of minor importance, has become a serious pest and an emerging problem particularly on *rabi* sorghum with the introduction of new genotypes of different maturity level in certain parts of Maharashtra, Andhra Pradesh, Karnataka and Tamil Nadu (Agrawal *et al.*, 1978). Both adults and nymphs suck the sap of young shoot and leaves, resulting in chlorosis. Macropterous (winged forms) and brachypterous (short winged forms) adults in either sex suck the sap of tender parts that lead to yellowing and stunted growth. Severe infestation at boot leaf stage results in twisting of top leaves ultimately preventing the emergence of ear head (Agrawal *et al.*, 1978). The bugs produce large amount of honeydew on which sooty mould (C.O: *Capnodium* spp.) developed affecting the photosynthesis. Besides this, the bug was reported to act as vector for sorghum stripe disease and maize mosaic disease virus (Peterschmitt *et al.*, 1991). As a result, crop loss to the extent of 41 per cent has been recorded (Hosamani & Chittapur, 1997).

MATERIAL & METHODS

The investigation on variations in biological parameters among the macropterous and brachypterous forms of *P. maidis* was undertaken at room temperature (28-30^{0}C) at Regional Agricultural Research Station, Raichur, Karnataka (16^{0}15^1 N Latitude, 77^{0}20^{1}E longitude and 398.4m AMSL) during 2003-04. Freshly emerged macropterous and brachypterous pairs from pure culture were allowed to mate in the laboratory in plastic clip cage (5 cm x 3 cm) with plastic lid on one side and muslin cloth with rubber band on the other side. The proximal end above the lid was placed on the upper surface of the leaf lamina (30 days old plant) by adjusting it in such a way that the midrib was in the centre of the cage. Observation on duration of the various developmental stages and adult longevity were recorded inside the cage. To study the nymphal development of macropterous and brachypterous forms, freshly emerged nymphs of macropterous and brachypterous forms were introduced separately into cages having potted plants of 10 days old. The soil surface below the plant was covered with black carbon paper to identify the fallen exuviae for recording moultings. Entire potted plant was covered with polythene sheet except the top, which was provided with muslin cloth for better ventilation for both nymph and plant.

RESULTS & DISCUSSION

Shorter incubation period of 7.27±0.64 days was noticed in eggs of brachypterous forms compared to macropterous form (8.35±1.05 days). However, there was no difference with respect to nymphal duration between brachypterous (16.25±1.86 days) and macropterous forms (16.06±2.03 days) (Table 1).

Table 1. Biological parameters of two forms of sorghum shoot bug

Life stages	Macropterous pair Duration (in days)		Brachypterous pair Duration (in days)	
	Range	Mean ± SD	Range	Mean ± SD
**Incubation period	6.50 – 9.50	8.35 ± 1.05	7.50 – 9.50	7.27 ± 0.65
***Nymphal development**				
I instar	2.00 – 3.75	3.35 ± 0.53	2.75 – 3.75	3.33 ± 0.38
II instar	2.25 – 3.25	2.85 ± 0.25	2.25 – 3.50	2.90 ± 0.46
III instar	2.25 – 3.50	3.17 ± 0.41	2.75 – 3.50	3.20 ± 0.26
IV instar	2.25 – 3.75	3.22 ± 0.58	2.25 – 3.75	3.27 ± 0.41
V instar	3.25 – 3.75	3.47 ± 0.26	2.25 – 3.75	3.27 ± 0.41
Total nymphal development	12.00 – 18.00	16.06 ± 2.03	13.00 – 18.50	16.25 ± 1.86
**Pre-mating (Hours)	6.50 – 11.50	8.10 ± 1.90	6.00 – 10.00	6.66 ± 1.60
**Mating (Hours)	0.53 – 1.36	0.61 ± 0.23	0.40 – 1.13	0.84 ± 0.21
**Pre-oviposition	0.75 – 3.75	2.40 ± 1.10	0.75 – 3.25	2.20 ± 0.90
**Oviposition	2.00 – 7.30	5.66 ± 1.74	3.66 – 8.00	5.93 ± 1.42
**Post-oviposition	15.45 – 24.91	17.87± 4.07	16.32– 22.78	18.95 ± 3.20
Adult longevity				
Male	7.50 – 18.00	10.40 + 4.49	4.50 – 16.50	8.95 + 4.42
Female	18.50 – 36.50	26.30 + 7.00	21.00 – 34.50	27.40 + 5.60
**Fecundity (No.)	22.00 – 72.00	48.90 + 18.50	16.00 – 106.0	67.00 + 30.20
Life Cycle (Egg to emergence)	18.50 – 27.50	24.41 + 3.08	20.50 – 28.00	23.52 + 2.51
Life cycle (Egg to complete death)	37.00 – 64.00	50.71 + 10.08	41.50 – 62.50	50.92 + 8.11

* Mean of 10 observations **Mean of 15 observations

The pre-mating period in brachypterous form ranged between 6.00 to 10.00 hrs with mean of 6.66±1.60 hrs, whereas, in macropterous form it ranged between 6.50 to 11.50 hrs with a mean of 8.10±1.90 hrs. The mating duration was found longer in brachypterous form (x 0.84±6.21 hrs.) as compared to macropterous form (x 0.61±0.23 hrs).

The mean pre-oviposition period and oviposition period in brachypterous and macropterous forms were 2.20±0.90 and 5.93±1.42 days, and 2.40±1.10 and 5.66±1.74 days, respectively. The hatching process in eggs laid by brachypterous forms took shorter duration (9-10 min). It was longer in macropterous forms (20-25 min). The adult longevity in macropterous male ranged between 7.5 and 18.0 with a mean of 10.40±4.4 days and in brachypterous form it ranged from 4.5 to 16.50 with an average of 8.95±4.42 days. Similarly, the adult longevity in macropterous female ranged between 18.5 to 36.5 with a mean of 26.3±7.0 days, and in brachypterous form it ranged between 21.0 and 34.5 with a mean of 27.4±5.6 days. This indicated that females lived longer than males in both the forms. The brachypterous forms recorded higher fecundity ranging from 16 to 106 eggs/female with a mean of 67.0±30.2 eggs/female. In macropterous forms fecundity ranged from 22 to 72 with a mean of 48.9±18.5 eggs/female. The total duration of life cycle was slightly more (24.41±3.08 days) in macropterous compared to brachypterous form (23.52±2.51 days). Due to these variations in life stages of brachypterous forms, they multiply faster and become dominant and cause severe damage. The present findings revealed a wide variation in biological parameters of the two forms. Earlier workers studied the biology of macropterous forms only (Catindig *et al.*, 1996). The variations exhibited by the shoot bug forms has practical implications and use in pest management

References

Agrawal, R. K., R. S. Verma. & G. S. Bharaj. 1978. Screening of sorghum lines against shoot bug. *JNKVV Res. J.* 12: 116.

Catindig, J. L. A., A. T. Barrion & J. A. Litsinger. 1996. Plant host range and life history of the corn Delphacid, *Peregrinus maidis* (Ashmead) (Hemiptera: Delphacidae). *Asia Life Sci.* **5:** 35-46.

Hosamani, M. M & B. M. Chittapur. 1997. Pest management in sorghum ecosystem In: *Sorghum Production Technology* (Ed.) S. M. Hosamani. University of Agricultural Sciences, Dharwad, Karnataka. 86pp.

Peterschmitt, M., A. S. Ratna, W. K. D.Sacks, V. R. Reddy & L. K. Mughogho. 1991. Occurrence of an isolate of maize stripe virus on sorghum in India. *Ann. of Appl. Biol.* 118: 57-70.

Sharma, H. C. 1985. Strategies for pest control in sorghum in India. *Trop. Pest. Manage.* 31: 167-185.

Swindale, L. D. 1990. *A Guide to Sorghum Breeding.* International Crop Research Institute for the Semi-Arid Tropics (ICRISAT), Patancheeru, Hydarabad. Andhra Pradesh.

61

Reproductive Behaviour of *Tetranychus neocaledonicus* Andre (Acari: Tetranychidae) Infesting *Moringa oleifera* Lam

SANGEETHA G. KAIMAL AND N. RAMANI

ABSTRACT

Mating behaviour of the vegetable mite, *Tetranychus neocaledonicus* (Andre) was studied in the laboratory at 34±1°C and 50±5% relative humidity using *Moringa oleifera* Lam as the host plant. Rearing of the life stages of the mite was carried out following leaf flotation technique. The process of sperm transfer was achieved through copulation immediately after the moulting of the female deutonymph. Time spent in copulation averaged 2 to 3 minutes. A male inseminated an average of 10-12 females during his lifetime, but the female copulated only once. Ovipositing females deposited eggs more frequently on the lower surface of the host leaves on webs constructed across the mid rib and major lateral veins. Specific site for oviposition was noted. *T. neocaledonicus* exhibited sexual as well as parthenogenetic mode of reproduction. Sex ratio was observed to be 1:10.

INTRODUCTION

The vegetable mite, *Tetranychus neocaledonicus* is a serious pest of over 110 different species of plants including flowers, fruits, vegetables, field crops, fodder crops etc (Sidhu & Singh, 1972; Jeppson *et al.*, 1975;

Puttaswamy *et al.*, 1976; Pande & Yadava, 1976; Puttaswamy & ChannaBasavanna, 1980; Ray & Rai, 1981; Sharma & Kushwaha, 1984; Manjunatha & Puttaswamy, 1989). *T. neocaledonicus* sucks the plant sap from plant cells producing white spots that gradually coalesce as feeding continues leading to acute chlorosis of the leaves. Affected leaves exhibited chlorophyll loss and subsequent drying up and shedding (Puttaswamy & Reddy, 1980). The present investigation was, therefore, undertaken to study the reproductive behaviour of a serious pest, *T. neocaledonicus* on a new so far unrecorded host plant, *Moringa oleifera* Lam, a plant well known for its multipurpose attributes, wide adaptability and ease of establishment. Furthermore, *M. oleifera* has high nutrient value, being rich in proteins, vitamins–A, B and C and minerals like calcium, iron and potassium (Freiberger *et al.*, 1998).

MATERIAL & METHODS

Studies on reproductive behaviour of *T. neocaledonicus* were initiated in an environmental growth chamber at 34±1°C and 50±5% relative humidity on fresh leaves of *Moringa oleifera* with petioles removed, excised from plants 3-5 days after expansion and kept in 120 mm petridishes lined with 110x110x5 mm cotton pads. The cotton pads were wetted daily with water to maintain the leaves' vitality. When the leaves showed signs of decay, all the mites were transferred to fresh ones. Five to ten colonies of active female deutonymphs were introduced and maintained until they became quiescent. Two to four new males were introduced into each culture set prior to initiation of moulting of the quiescent deutonymphs. Studies on parthenogenesis were initiated starting from 5-10 quiescent female deutonymphs which moulted into adult females and laid their first batch of eggs in the absence of males. Adult mites of each colony were transferred to fresh leaves every 24 hrs. Regular observations were made on each culture set at 6h intervals under 32 x magnifications, in order to gather information on mating and oviposition.

RESULTS & DISCUSSION

Mating: *T. neocaledonicus* exhibited marked sexual dimorphism with the narrow wedge shaped and pointed hysterosoma for the male and a large, oval- rounded hysterosoma for the female (Plate I, Fig. 1). Mating occurred immediately after the moulting of the female deutonymph. In the laboratory cultures, males emerged comparatively earlier than the females. Soon after emergence, they were found moving in search of female deutonymphs in quiescence. As soon as a male encountered a female deutonymph ready for moulting, the male rested close to the female, with his forelegs extending over the dorsum of the latter (Plate I, Fig. 2). When a split appeared on the cuticle of the female, the male was found pulling the posterior half of the cuticle and helping the female to come out of the exuvium (Plate I, Figs. 3 & 4). On several occasions, mating occurred even before completion of the moulting of the female deutonymph. During copulation the male crawled beneath the female hysterosoma, lifting up the posterior end of the latter. Meanwhile, the male gripped himself firmly on the leaf surface by means of his hind legs. Posterior tip of the male hysterosoma was then held in a characteristic curved position, protruding the club shaped aedeagus upwards to the vagina of the female (Plate I, Figs. 5, 6 & 7). The mating lasted for 2 to 3 minutes. At the end of the process, the male retracted the aedeagus and moved backwards. Then he moved away in search of a new female. A single male was found inseminating an average of 10-12 females during his lifetime. The female copulated only once in her lifetime.

Parthenogenetic versus sexual development: *T. neocaledonicus* performed both sexual and parthenogenetic reproduction. The process of embryonic and postembryonic development was similar in both types of reproduction. However, all the progeny were found to be males in the case of parthenogenetic development while in the sexual reproduction both males and females were produced with a sex ratio of 1:10 (Manjunatha & Puttaswamy, 1989). Slight difference was noted in the duration of the developmental stages in both the types of reproduction. The total duration of development was comparatively short in the case of parthenogenetic progeny.

Oviposition: Adult females were found depositing eggs more frequently on the lower surface of the leaves of the host plant, *M. oleifera* and specific selection of site was noted during oviposition. However, in rare instances, eggs were laid on the upper surface also. Ovipositing females were found constructing silken web produced by a pair of glands located near their mouth, across the midrib and major lateral veins prior to the deposition of the eggs. This webbing activity was seen to coincide with oviposition also. During oviposition, the females after making a firm grip on the leaf surface with her first two pairs of legs, slightly lowered the hysterosoma for extruding the egg. Subsequently, the female was found moving away from the site. A tendency of depositing the eggs in close proximity was observed and hence the eggs deposited were usually found in colonies (Plate I, Fig. 8).

Plate 1

F- Female, M- Male & Eggs of *Tetranychus neocaledonicus* on *Moringa oleifera*

CONCLUSION

In *T. neocaledonicus*, males emerged comparatively earlier than the females, approached the quiescent female deutonymphs and copulated with the female immediately after the emergence of the latter. These observations are in agreement with the behavioural activities observed in the case of *T. urticae* (Penman & Cone, 1972), *T. evansi* (Qureshi *et al.*, 1969), *Panonychus citri* (Beavers & Hampton, 1971) and *Eutetranychus orientalis* (Banu & ChannaBasavanna, 1972). Hence it may be concluded that almost all the tetranychids known so far follow a basic pattern of mating. Further, the tendency for deposition of the eggs adjacent to the mid rib of the leaves of *M. oleifera* on the lower surface was also recorded in several other tetranychids on their respective host leaves. This would definitely ensure better protection to the mite from desiccation as well as washing effects of rain when compared to the upper surface, thereby enhancing its survival rate considerably.

References

Banu, K. & G. P. Channabasavanna. 1972. Plant feeding mites of India: A preliminary account of the biology of the spider mite *Eutetranychusorientalis* (Klein) (Acari: Tetranychidae). *Mysore J. of Agri. Sci.* 6(3): 253-268.

Beavers, J. B. & R. B. Hampton. 1971. Growth, development and mating behaviour of the citrus red mite (Acari: Tetranychidae). *Ann. Entom. Soc. Amer.* 64: 804-806.

Freiberger, C. E., D. J. Vanderjagt, A. Pastuszyn, R. S. Glew, G. Mounkaila, M. Millson. & R. H. Glew. 1998. Nutrient content of the edible leaves of seven wild plants from Niger. *Pl. Foods for Hum. Nutr.* 53(1): 57-69.

Jeppson, L. R., H. H Keifer & E. W. Baker. 1975. In: *Mites injurious to economic plants*. University of California Press, Berkley, Los Angeles & London. 614pp.

Manjunatha, M. & Puttaswamy. 1989. Life history of *Tetranychus neocaledonicus* under green house conditions. *J. Acarol.* 11(1 & 2): 35-40.

Penman, D. R. & W. W. Cone. 1972. Behaviour of two spotted male spider mites in response to quiescent female deutonymphs and to web. *Ann. Entom. Soc. Amer.* 65: 1289-1293.

Puttaswamy & G. P. Channabasavanna. 1980. Competition between *Tetranychus neocaledonicus* and *T. ludeni* (Acari: Tetranychidae). *Ind. J. Acarol.* 4 (2): 49-54.

Puttaswamy & D. N. R. Reddy. 1980. *Tetranychus neocaledonicus* (Andre) (Acari: Tetranychidae), a new pest on cardamom. *Acarol. Newsletter* 10: 6-7.

Puttaswamy, M. C. Devaiah, & H. R. Rangaswamy. 1976. A new record of vegetable mite, *Tetranychus cucurbitae* (Rahman and Sapra). *Curr. Res.* 45: 117-118.

Qureshi, A. H., E. R. Oatman, & C. A. Fleschner. 1969. Biology of the spider mite, *Tetranychus evansi* Pritchard and Baker. *Ann. Entom. Soc. Amer.* 62: 898-903.

Ray & Rai. 1981. Biology and Control of *Tetranychus neocaledonicus* (Acari: Tetranychidae) on lady's finger at Varanasi.. In: *Proc. Symp. on Acarol.* (Ed.). G. P. ChannaBasavanna. Acarological Society of India, University of Agricultural Sciences, Bangalore. pp.41-46.

Sharma, A. & K. S. Kushwaha. 1984. Susceptibility of different varieties of brinjal to *Tetranychus neocaledonicus* (Acari: Tetranychidae). *Ind. J. Acarol.* 6(2): 100-103.

Sidhu, A. S. & G. Singh 1972. Laboratory studies on the host range of vegetable mite, *Tetranychus cucurbitae* (Rahman and Sapra). *J. Res. Punjab Agricultural University.* 9 : 313-314.

62

Biology, Adult Emergence and Mating Behaviour of Melon Fly, *Bactrocera Cucurbitae coquillett* (Diptera: Tephritidae)

*SHIVAYYA, V., C.T. ASHOK KUMAR
AND A.H. JAYAPPA

ABSTRACT

Studies on biology of melon fly, *Bactrocera cucurbitae* Coq. was carried out during 2003 and 2004 in Bangalore, Karnataka, India. The results of the investigations revealed that the maximum number of adult emergence occurred between 09.00-11.00 hours. Flies mated during the night between 18.30-01.30 hours. Ovipositional activity started a day after mating. The fecundity ranged from 188 to 250 eggs per female. It took 26-28 hours, 4.5-7.5 days and 7.00-11.50 days for incubation, larval and pupal period, respectively. The total life cycle of melon fly varied from 12.50, 16.50 and 20.50 days during December-January, March-April and August-September, respectively. The flies survived for 18.4 ± 0.64 days with only sugar and longevity was enhanced to 27.2 ± 2.8, 34.2 ± 3.8 days when fed with honey + water, Sugar + water and sugar + yeast + peptone + water, respectively. Non-supply of water led to sudden death of the flies. Flies did not show the reproductive maturity, mating and oviposition activity but lived for 35 days when only sugar and water were provided.

INTRODUCTION

Bitter gourd is attacked by several species of insect pests of which, melon fly, *Bactrocera cucurbitae* causes serious damage to fruits. The females puncture the soft and tender fruits with ovipositor and lay eggs below the epidermis of fruits. The maggots feed on the pulp. The puncture may also serve as a site of secondary infection by pathogens causing rottening and pre mature falling. *B. cucurbitae* attack in bitter gourd can cause damage varying from 15 to 100 per cent (Gavigowda, 1976). It is a known fact that a source of carbohydrates and proteins is necessary for normal growth and development of insects. Work on these aspects is very limited in phytophagous dipterans. Looking at the apparent importance of the pest to bitter gourd, and the lack of information pertaining to the biology of *Bactrocera cucurbitae* in India, the biology of the melon fly was studied at Bangalore, Karnataka, South India.

MATERIAL & METHODS

Bitter gourd, *Momardica charantia* L plants were maintained throughout the study period to provide continuous fruits to melon flies under greenhouse conditions at Department of Entomology, UAS, GKVK, Bangalore. The damaged fruits of bitter gourd along with maggots of melon fly were collected from the fields and were reared in laboratory to obtain pure culture and also to study the developmental behaviour.

The infested fruits were kept in a plastic container (20 x 20 x 8 cm) on 5 cm moist thick layer of sieved moist sand to facilitate pupation, which was placed inside the rearing cage (35 x 30 x 35 cm). The rearing cage was designed to have wire mesh wall on three sides, glass wall at the top and wooden door at one side. A round trap door was provided in the wooden door to facilitate collection of adult flies for experimental purpose and also to provide food and water. Freshly emerged male and female flies were transferred to another rearing cage and provided with uninfected, tender bitter gourd fruit discs. The bitter gourd fruit discs of 1 cm thickness served as oviposition substrata and food for maggots. The entire fruit culture was maintained at mean temperature of 26 ± 1°C with mean relative humidity of 60 ± 5.0 per cent. Immediately after emergence the flies were fed on the following materials for studying the reproductive maturity, mating, fecundity and longevity of adult flies. The experimental set up consisted of six treatments and three replications as given below:

T1 = Water only
(water was provided through sterilized cotton soaked in water which was kept in petri plate)
T2 = Sugar only (Sugar was provided in petri plate)
T3 = Sugar and water (1:1)
T4 = Sugar, peptone, yeast and water (2:2:1:2 ratio)
T5 = Honey and water (1:5 ratio)
T6 = No food (neither water nor food)

Distilled water was provided separately through soaked sterilized cotton wool placed in petri plate (6 cm diameter) for all the treatments. Food and water was replaced once in two days. Observations on biology of the fly were recorded at regular intervals. In each treatment 5 pairs of flies (male and female) were studied.

RESULTS & DISCUSSION

Egg: Female *B. cucurbitae* possess bristles at the end of the ovipositor, which apparently serves as tactile organs to detect suitable place to deposit eggs. While laying eggs the female bends its abdomen to the right of the body and inserts its ovipositor by back and forward movements of the abdomen. It forms a cavity about 2-4 mm deep in the fruit and makes the wall of the cavity impervious to moisture by a gummy secretion so as to protect the eggs from excess moisture. Freshly laid eggs were glittering, white, cylindrical, slightly curved and more convex on the dorsal side. Eggs measured 1.57 mm in length and 0.34 mm in width.

Eggs hatched in 16-18 minutes. Before emergence the maggot inside the egg showed vigorous movements of its mouth, hooks and pharyngeal skeleton, which lead to rupturing of anterior portion of egg shell. The maggots emerged out of the egg and started feeding on the host. The egg duration varied from 26 to 38 hours, (26 hours during March-April, 32 hours during August-September and 38 hours during December-January). These results agree with Back & Pemberton (1914). Vargas *et al.*, (1984) and Tewatia (1997). Vishva (2005) also reported incubation period of 1-9 days depending upon temperature and relative humidity.

Maggot: The apodus maggot had three instars. Freshly emerged first instar maggots were white, with black mouth hooks and two white thread-like respiratory tracks on its back. The spiracles were prominent. The first instar lasted 28.09 ± 0.34 hours and measured

2.18mm long and 0.35mm wide. Second instar maggot lasted 24.00 hours and was 6.15mm in length and 1.75mm wide. The posterior part of the maggot was crimson and anterior part white. The present findings are in accordance with Vishva's observations (2005).

The full-grown maggot exhibited a peculiar mode of locomotion by quick spring like jumps. The jumping drew the insect towards darkness, which facilitated in finding a suitable location for pupation. The full-grown maggot pupated usually inside the soil at 2.5 to 10cm depth. At times pupation occurred inside the fruit itself. Maggot preferred moist soil for pupation, which lasted 13 to 20 hours. The maggot period was 7.50, 4.50 and 5.50 days during December-January, March-April and August-September, respectively (Table 1).

Pupa: The puparium was cylindrical or elliptical, 4 to 6mm long and 1.7 to 2.5mm wide brownish yellow with eleven distinct segments. The first segment of puparium was composed of first and second larval segments. The pupal period was 11.50, 7.00 and 10.75 days during December-January, March-April and August-September, respectively (Table 1). The present findings are in accordance with studies of Narayanan & Batra (1960), Gavigowda (1976) and Vishva (2005).

Adult: The peak emergence of adult melon fly was observed between 09.00 to 10.00 hours and continued up to 16.0 hours. The freshly emerged adult flies were pale yellow and appeared spider like without any marking on the thorax. After few minutes, it spread wings and developed colour on the wings and thorax. Flies took 25-35 minutes to gain a fly look. These observations are in agreement with Gavigowda (1976), but Narayanan & Batra (1960) reported peak emergence of adults between 6.00 hours and 8.00 hours in summer, 9.00 hours and 11.00 hours in winter. Rahman *et al.*, (2004) reported peak emergence of adult between 5.00 and 5.30 hours, just before sun rise.

Mating: The excited male was observed vibrating wings and moving towards the adult female flies by jumping on them till he got an excited female. The male suddenly jumped on the excited female, rested on the back of the female for copulation by holding female firmly with his pair of legs. The copulation lasted from a minimum of 2.00 hours to maximum of 6.50 hours (Table 3). Copulation spanning 1-7 hrs was recorded by Vishwa (2005). Sayed (1972) and Vishva (2005), reported that the diminishing light condition induced mating in the fruit flies. Bateman (1972) also reported that falling light at dusk acted as a stimulus for initiation of sexual activity in many species of Tephritidae including *Bactrocera cucurbitae*.

Fecundity: The fecundity of a female in laboratory varied from 188 to 250 eggs with an average of 222.80 eggs (N= 100). Similar observations were recorded by Sayed (1972). The results obtained on oviposition behaviour and fecundity are in close accordance with the results of earlier workers who reared flies on different kind of foods *viz.*, Sugar + yeast + peptone + water (Gavigowda 1976), water + banana + protein + sugar (Vishva, 2005).

Total life cycle (Table 2): Adults died within 1.0 to 1.5 days with an average of 1.20 days, when they were provided neither food nor water immediately after their emergence. The longevity extended up to 3.5 days by providing only water. When flies were provided with only sugar, they lived up to 21 days. By supplying sugar and water they lived up to 35 days. Whereas they lived for 145 to 150 days when provided sugar + peptone + yeast + water and for 61-63 days when provided with honey and water.

Longevity of adults varied with the type of food. Addition of carbohydrate improved very much the longevity of adult flies (Table 2). Gavigowda (1976) also reported that the flies survived for 17 days with only sucrose and longevity was enhanced to 32 days with sugar and water. The total life cycle of melon fly from egg to adult varied from 12.50, 16.50 and 20.50 days during December-January, March-April and August-September, respectively under laboratory in Bangalore (Table 1).

Table 1. Duration of immature stages of *B. cucurbitae* on bitter gourd during 2003-04

Period of Study	Duration of stages						Total life cycle (days)	
	Egg		Larva (Days)		Pupa (Days)			
	Range	Average	Range	Average	Range	Average	Range	Average
Dec-Jan	37.5-38.5	38.0	7.2-7.7	7.5	10.7-2.2	11.5	19.2-21.8	20.5
Mar - Apr	24.7-27.3	26.0	3.7-5.2	4.5	6.8-7.2	7.0	10.5-14.5	12.5
Aug-Sep	31.5-32.5	32.0	4.7-6.2	5.5	8.2-11.2	10.7	15.2-17.2	9.5

Table 2. Longevity of adult melon flies on different food sources

Food sources	Numbers tested		Adult longevity (days)	
	Female	Male	Range	Mean ± SD
Water	5	5	2.0-3.5	2.80±0.70
Sugar	5	5	18.0-21.0	18.30±2.60
Sugar + water	5	5	33.0-35.0	34.20±0.80
Sugar + peptone + yeast + water	5	5	145.0-150.0	148.0±2.00
Honey + water	5	5	25.0-30.0	27.2±2.80
No food or water	5	5	1.0-1.5	1.20±0.30

Table 3. Duration of mating in *B. cucurbitae*

Pair No.	From (Hrs.)	To (Hrs.)	Total mating period (Hrs.)
1	19.20	23.10	3.50
2	19.30	21.30	2.00
3	18.30	00.30	6.00
4	18.45	23.15	4.30
5	19.00	21.10	2.10
6	18.30	22.45	4.15
7	19.20	00.45	5.25
8	18.45	23.45	5.00
9	19.30	01.20	5.50
10	20.15	01.20	5.05
Mean	-	-	4.55

References

Back, E. A. & C. E Pemberton. 1917. *The Melon fly in Hawaii.* United States Department of Agriculture 491: 1-64.

Bateman, M. A. 1972. The ecology of fruit flies. *Ann. Rev. Entom.* 17: 493-518.

Gavigowda, M. 1976. *Studies on fruit flies infesting cucurbits with special reference to bionomics of Dacus cucurbitae (Coq.) (Diptera: Tephrtidae) and possibility of utilizing susceptible muskmelon cucurbits from its attack.* M.Sc. thesis, Submitted to University of Agricultural Sciences, Bangalore. 115pp.

Narayanan, E. S. & H.N. Batra. 1960. *Fruit flies and their control.* A monograph by Indian Council of Agricultural Research, New Delhi. 66pp

Rahman, A. K., M. Z. Alam., M.N. Islam. & M.M. Hossain. 2004. Studies on the behaviour of adult fruit fly (*Bactrocera cucurbitae*) after emerging from pupae. *Bull. Inst. Trop. Agri.* Kyushu University. 26: 1-9.

Sayed, R. A. 1972. *Studies on tephritids and their natural enemies in West Pakistan: Dacus cucurbitae Coq. Tech. Bull. Common Wealth Insti.Biol. Con.* 14: 63-75.

Tewatia, A. S. 1997. Melon fruit fly resistance in cucurbits: A review. *Vegetable Science.* 24: 79-82.

Vargas, R. R., P. Miyashita. & J. Nishida. 1984. Life history and demographic parameters of three laboratory-reared tephritids. *Ann. Entoml. Soc. America* 7: 651-656.

Vishva, N. S. 2005. *Studies on biology, behaviour and management of melon fruit fly, Bactrocera cucurbitae (Coq.) (Diptera: Tephritidae).* M.Sc. Thesis submitted to University of Agricultural Sciences., Bangalore. 122 pp.

63

Mating and Ovipositional Behavior of the Indian Uji Fly, *Exorista bombycis* (Louis), A Parasitoid of the Mulberry Silkworm *Bombyx mori*

FATIMA. S., C. SIDDAPAJI
K.P. CHINNASWAMY AND K. RASHMI

ABSTRACT

The Indian Uji fly, *Exorista bombycis* (Louis) is the lone insect parasitoid of silkworm *Bombyx mori* L. causing loss to growers of silk cocoon. The behavioral study was undertaken to understand the mating and oviposition of the parasitoid. The fly infests silkworm by laying eggs on silkworm and subsequently upon hatching, maggot gains entry into system. The adult flies of either sex did not exhibit characteristic courtship behavior. The pre-mating duration was slightly high during Dec-Jan (19.52$\pm$ 6.8h) compared to April-May (17.0$\pm$ 6.1h). The mating duration also varied similarly. The male Uji flies were found ready for mating earlier to females. Usually, a pair of Uji flies mate twice in 24 hours. A single male could mate with a maximum of three females. On the other hand a single female accepted two males during its lifetime. Studies established that the Uji fly *E. bombycis* exhibits both polygamous and polyandrous behaviour.

The gravid Uji fly scans the host with its antennae, chooses healthy hosts and walks over unsuitable worms. Thus female Uji fly searches for the host discriminately. The pre-ovipositional period ranged from 39.9h to 51.2h. The oviposition period ranged from 6 to 22 days. Mated female Uji flies were observed to oviposit till death. Maximum number of eggs were deposited between 3rd and 6th day. Virgin females laid sterile eggs (168), did not glue to silkworm, and stuck to finger on touching, whereas mated females laid more number of eggs that were fertile implying that mating stimulated oviposition. The duration of mating had no influence on fecundity but had positive significant effect on egg hatchability. Mating and oviposition behavour could be exploited in Integrated Pest Management of the fly parasatoid.

INTRODUCTION

The Indian uji fly, *Exorista bombycis* is the lone insect parasitoid of the mulberry silkworm, *Bombyx mori* with potential to cause losses of serious consequences to the sericulture industry. The mulberry silkworm when infested by this pest generally continued to feed as before, and in some cases worm spun a cocoon, but it invariably perished when the maggots cut their way out. The cocoon made by the parasitized worms was generally poor one and was unfit for effective reeling because of the hole made by the maggot while escaping from it. The uji fly does not reproduce without mating. It is this mating process and subsequent hatching of the eggs, which causes all the damage to its host. Fertilized eggs hatch and unfertile eggs laid by unmated virgin females do not hatch, hence there is no parthenogenesis. Owing to the fact that insect reproduction usually involves mating between the two sexes and considering the importance of mating on reproduction, the present study was undertaken. If mating of Uji fly can be disrupted at an appropriate time, it's damage to silk production can be completely controlled.

MATERIAL & METHODS

The pre-mating duration was recorded as the time from adult emergence to initiation of first mating in each case of the enclosed individual pairs and observations recorded.

As the uji flies began to emerge from the mass culture, they were transferred to glass jars age wise and closed with muslin cloth on the rim. Adults of particular age class were combined and the males and female pairs were enclosed in separate glass jars, covered with muslin cloth. Ten percent glucose solution was provided as diet. The adults mated within a couple of hours inside the enclosure and were recorded as the first mated pairs for that particular age class. The pairs were allowed to mate without causing disturbance and the time in coitus position was recorded as one mating duration till their natural depairing.

To study whether the males are polygamous, one unmated male was enclosed with three unmated females, two marked differently and one unmarked, in a jar separately covered on the top with thin muslin cloth. The experiments were maintained with adequate replications and checks were maintained. To find out if females were polyandrous, one female was enclosed with three males, two marked and one unmarked, soon after emergence in a glass jar, covered with muslin cloth at the rim and the experiment was replicated four times, and number of matings and duration of each mating was recorded.

Pre-ovipositionsal period was recorded as the period from the time of emergence of the female uji fly to the deposition of the first egg on the silkworm.

To study the number of eggs laid per female, 10 pairs of newly emerged uji flies were enclosed in ten glass jars with 5 fourth instar silkworms with mulberry leaves provided in each jar each day. The silkworms were replaced once every day, till the female uji fly died, and the eggs glued to and black scar formed on silkworms were counted each day to record the number of eggs laid and hatched per day. The percentage of egg hatchability was calculated as

$$\frac{\text{No. of black scars x 100}}{\text{No. of eggs}}$$

To record the ovipositional pattern of mated and virgin females, a pair (one female and one male) of uji flies and a just emerged female were enclosed separately in 10 replications in a glass jar soon after emergence and provided 10 fourth and fifth instar PM x NB_4D_2 silkworms each day, till the uji flies died. Fecundity and egg hatchability were recorded as mentioned earlier.

In order to study the effect of duration of mating on the number of eggs laid and subsequent hatchability, three of first mated pairs were kept under observation from initiation of mating to 30 minutes of mating when they were depaired and separated. The females were enclosed in a glass jar and 5 early fifth instar PM x NB_4D_2 worms were offered for oviposition. The worms were replaced daily once. Fecundity and egg hatchability were recorded. The above procedure was followed meticulously when pairs naturally depaired after 80, 163 and 192 minutes of mating. To study the effect of repeated mating 5 pairs were maintained continuously from mating to oviposition.

RESULTS

*Mating Behavior:*The male mounted the female without any prelude. The unwilling female did not respond and flew away with a characteristic buzz. Many a times the males were observed chasing the females and even mounting them when a large number of uji flies were confined in the insect rearing cages.

Copulatory Behavior : Copulation occurred after the male had gripped the female at the first abdominal segment with the pro-thoracic and meso-thoracic legs while inserting its aedugus into the genital orifice of the female. The male soon slided a little forward to hold the female's head and abdomen with it's fore and mid legs respectively, while the hind legs rested on the substratum.

Pre-mating duration: The uji flies mated only after a minimum lapse of 8.16 hours after emergence in November-December and 8.60 hrs in April–May.Though the pre-mating duration was more in Nov-Dec (19.5$\pm$ 6.8h) than in April-May (17.10$\pm$ 6.1h) the difference was not significant.

Mating duration: Mating duration of the uji flies was also slightly more in November-December (2.46$\pm$0.7h) than in April-May (1.85$\pm$ 0.6h). But the difference was not significant.

Age related sequence of mating success: In April-May, maximum number of uji flies mated when three day old females were enclosed with two day old males (38.9%) followed by adults of equal age, two day old males and females (28.20%).

In the combination of one-day-old female with one day old male, two day old female with one day old male and, three day old female with four day old male and, three day old female with four day old male, only 12.5% mating pairs were found.

In November-December mating success was 32.25%, 28.79%, 19.00%, 17.4%, 11.47%, 9.67% and 4.87% in the combination of three day old female with two day old male, two day old female with three day old male, equal aged male and female (two days) followed by three days old, two day old female with two day old male, three day old female with four day old male, four day old female with three day old male and four day old female with four day old male respectively. No mating was observed in one day old and one day old female combination.

In April-May males mated between one day and four days of age, while females mated between one day and three days of age. However in November to December, female uji flies mated between the ages of one and five days.

Multiple mating behavior: In order to find out if uji fly males are polygamous, two marked females and one unmarked female were confined in a glass jar, the male-mated first with the unmarked female for 100 min. After a resting period of 120 min, it again mated with one of the marked females for another 33 min. The same pair continued to mate for another 97 min, after a break of 5 min. In another replication, the male mated with the marked female for 26 min, after a break of 11 min. It again mated for the second time with the same female and did not mate any more for the rest of its life.

In the third replication, the male mated only once with the unmarked female for 192 min. and did not mate again till death.

In the last replication, the male mated with a marked female for 74 min. after a resting period of 7 min, mated with the unmarked female for 142 min, rested again for 142 min, rested again for 205 min and mated with the second marked female for 126 min.

To find out if females are polyandrous, one female was enclosed with three males (two marked and one unmarked); the female mated with one of the marked males continuously for 204 min and did not allow any other male to mate with it for the rest of its life.

In the second replication the female mated with the unmarked male for 36 min, rested for two min. and mated again with a marked male for 147 min.

In the third replication, the female mated with the marked male for 89 min and after resting for 43 min. mated again with another marked male for 163 min.

In the fourth replication, the female mated for 5 min with one marked male and again after a break of 3 min. mated with the unmarked male for 193 min.

Ovipositional behavior: The female Indian uji fly after mating searches for the host discriminately and not randomly. The uji fly has been found scanning the host with her antenna. If the host is unsuitable i.e., if it already has uji eggs or is diseased or small, she walks over it and searches for a new host (silkworms in trays). Thus the uji fly exhibits a lot of maternal care while ovipositing.

Pre-ovipositional and ovipositional period: The mean pre-ovipositional period was 44.08 $\pm$ 5.5 h, and the mean ovipositional period for mated uji flies (n=10) was 14.1$\pm$ 4.6 days.

Ovipositional Pattern: Mated female uji flies continued to oviposit till they died (Table-6). The mean number of eggs laid per female was 501$\pm$108 (range 305 –731). Maximum number of eggs laid was between third and sixth day of oviposition.

A gradual decline in the number of eggs laid each day was noticed from the seventh day onwards. A max of 111 eggs was laid by a single fly in 24 hours (Table 6). On comparing the daily ovipositional pattern of virgin and mated uji flies, it was found that virgin uji flies started egg deposition on the second day of emergence, while mated uji flies began oviposition on the third day of emergence.

The ovipositional period for virgin females was only 10 days, while the same for mated females was 14 days. The eggs laid by virgin females were straw colored and did not stick to the host body (substratum) and were laid in clumps.

The virgin females laid less (168) number of eggs than the mated ones (463).The average egg hatchability in the case of mated females was 75.7 percent, and none for virgin females.

DISCUSSION

Mating behaviour: The Indian uji fly, *Exorista bombycis* did not exhibit characteristic courtship behaviour although Veeranna & Nirmala (1993) and Narayanaswamy (1991) reported elaborate courtship behavior. The present findings are in conformity with those of Kasturbai *et al.*, (1986), who reported that male flies, when attracted to the female exhibited typical mating behavior like wing fluttering and abdominal lifting. Siddapaji (1985) reported that copulation occurred with the male on top of the female. The mating partners made a characteristic buzzing sound as they approached each other and entered into coitus, and when disturbed took to wings (Cotes, 1989).

The pre-mating duration of the uji flies was slightly more in December-January (19.52 ±6.8h) than in April-May (17.0± 6.1h). There is wide variation in pre-mating duration in different regions such as 4h from Berhampore (Sriharan *et al.*, 1971), 1-3 days in Japan (Anon 1975), 2 days and 18h in March–April (Siddapaji, 1985), and 10h to 3days in Bangalore (Veeranna & Nirmala, 1989) The difference could be due to differences in handling, and fluctuations in climatic conditions prevailing during the different parts of the year.

The mating duration also varied; it was more in December-January (2.6±0.7) than in April-May (1.85±0.6h). Several workers have reported various durations such as 25-147 min (Sriharan *et al.*, 1971), average of 1h 42 min (Datta & Mukhrjee, 1978), 10-47 min (Siddappaji, 1985), 50-129 min (Narayanaswamy, 1991) and 1.34-1.69 h (Veeranna & Prasad, 1993).

The present study establishes that the male uji flies were ready for mating before the females of the same age. Mating between younger males and older females was a common feature in the present study and is reported for the first time

In the present study a single pair of uji flies were found mating two times in 24 hr, a single male was capable of mating with a maximum of 3 females, while a single female was capable of mating 2 times during its life time, thus establishing both polyandrous and polygynous behaviour in *E.bombycis*. The reports that mating occoured 3 times in 24 hr and 4 times in 24 hr is in agreement with the present study. In Japan the flies mated several times during forenoon (Anon, 1975), as confirmed by repeated mating in the current study. Jolly (1981) reported females as polyandrous, mating 2-4 times in 24 hr. Siddapaji (1985) observed both polyandry and polygyny by adopting MRC technique in the open. Narayanaswamy (1991) has also reported polygamy in *E. bombycis* in the laboratory.

Ovipositional behaviour: The female Indian uji fly searches for the host discriminately. The uji fly scans the host with its antenna, chooses healthy worms and walks over unsuitable worms (diseased, unhealthy, small etc.), thus uji fly exercises great maternal care in selection of worms/aspects of time in gluing the egg to inter-segmental region. Such behaviour was reported earlier by Siddapaji (1985). The uji fly was found to induce stretching of the body segments through ticklish movements, and secreting a quick drying gum on the under surface of the egg and depositing it perpendicular to the plane. According to Siddappaji (1985) the female uji fly also leaves a kind of odor which repels others from ovipositing on the same worm which phenomenon is partly responsible for the solitary parasitization mechanism.

The pre-ovipositional duration for the Indian uji fly, ranged from 33.9h to 51.2h with mean of 44.08± 5.5h. In Japan the pre-ovipositional period of this fly was 24-144h, and in India, it was 1.04±1.9h days (Datta and Mukherjee,1978), 6-48h (Siddappaji,1985) and 37.9to 52.3h (Veeranna & Nirmala, 1989).

The ovipositional period of the *E. bombycis* ranged from from 6-22 days with a mean of 14.1±4.6days, the longest ovipositional period under Berhampore conditions was 20 days (Sriharan *et al.*, 1971), 7-8 days(Jolly, 1981) and 8.12-12.36 days in Bangalore (Narayanaswamy, 1991).

Mated females of the Indian uji fly were observed to oviposit till death. Maximum number of eggs was deposited between 3-6th days of oviposition. A great decline in the number of eggs laid on each day was noticed from 7th day onwards. These observations more or less agree with those of Sriharan *et al.*, (1971) who reported that the female uji flies oviposit till they were alive and most eggs were laid between 4-7 days. Patil & Govindan (1984) reported maximum egg lying on 5th day after mating; however more number of eggs were laid between 2nd to 6th days of copulation and reduced thereafter. Narayanaswamy (1991) recorded maximum fecundity on 5th day. The discrepancy in respect of the oviposition period could be due to climatological variations.

Fecundity: In the present study, mean number of eggs laid by a gravid uji fly was 501±108 (305-731) eggs. Several workers have reported a wide variation in fecundity.

The unmated (virgin) females laid sterile eggs. A comparison between the daily ovipositional pattern of virgins and mated Indian uji fly indicate that mating somehow stimulated oviposition. A longer duration was required for selecting suitable site for oviposition, and the female selected silkworms with great maternal care. Virgins laid less number of eggs (168 sterile eggs) that did not hatch and the mated females laid more number of eggs that were fertile (Average, 436.00

eggs) that hatched, implying that mating increased fecundity. There have been no such studies reported earlier.

The duration of mating of the uji flies had no influence on the fecundity. The variation whatever noticed was not of any significance, thus agreeing with the report of Veeranna & Prasad (1993). The duration of mating had a significant positive effect on egg hatchability, in that, longer the duration of mating greater the egg- hatchability at each installment. These observations contradict the findings of Veeranna & Prasad (1993) who reported that the duration and frequency of mating of the uji flies did not have any influence on egg hatchability.

CONCLUSION

The Indian uji fly did not exhibit characteristic courtship behavior. The pre-mating duration was more in December-January (19.52± 6.8h) than in April- May (17.0± 0.6h), while mating duration was also more in December-January (2.46± 0.7) than in April-May (1.85± 0.6h). The present study established that the male uji flies of the same age were ready for mating earlier than females and that the Indian uji flies are both polygamous and polyandrous.

The mean pre-ovipositional period for the Indian uji flies was 44.08± 4.6days. Mated females were observed to oviposit till death. Maximum number of eggs laid by a single gravid uji fly was 501±108. Virgin females laid less number of eggs (168) that were sterile, while mated females laid more number of eggs that were fertile (436), implying that mating increased fecundity.

References

Anon 1975. *Text Book of tropical Sericulture.* Japan overseas Volunteers, Tokyo, Japan, pp.540-547.

Cotes, E. C. 1889. An account of the insects and mites that attack the tea plant in India. In: *The tea Insects of India. Indian Mus Notes* 3: 9-22.

Datta, R. K. & P. K. Mukherjee. 1978. Life History of *Tricholyga bombycis* (Diptera: Tachinidae), a parasite of *Bombyx mori* (Lepidoptera: Bombycidae) *Ann. of the Ent. Soc. America* 73: 767-770.

Jameson, A. P. 1922. *Report on the diseases of silkworm in India.* Government Printing Press, India. Pp.62-64.

Kasturibai, A. R., D. Mahadevappa, M. R. Nirmala. & H. K. Jyothi. 1986.Control of uji flies by semio- chemicals. *Curr.Sci.* 55: 1038-1039.

Narayanaswamy, K. C. 1991. *Population Dyanamics of Indian uji fly, Exorista bombycis(Louis) (Diptera: Tachinidae).* Ph.D.Thesis, University Agricultural Sciences, Bangalore. 223pp.

Patil & R. Govindan. 1984. Biology of uji fly, *Exorista sorbillans* (Wiedemann)(Diptera: Tachinidae) in silkworm *Bombyx mori. Indian J. Seric.* 23:22-31.

Siddapaji, C. 1985. *Bioecology and Management of the Indian uji fly, Exorista sorbillans (Diptera: Tachinidae), a parasite of mulberry silkworm.* Ph.D. thesis, University Agricultural Sciences, Bangalore. 166pp.

Sriharan, T, P., M. V. Samson., S. Krishnaswami. & R. K. Datta. 1971. Laboratory investigation on uji fly, *Tricholyga bombycis* Beck. a tachinid parasite of silkworm *Bombyx mori* L. *Indian J. Seric.* 10(1): 14-22.

Veeranna, G. & M. R. Nirmala. 1989. Courtship and mating behavior of the uji fly, *Tricholyga bombycis* Beck. (Diptera: Tachinidae). *Entomon.*14: 85-89.

Veeranna, G. & N. R. Prasad. 1993. Reproductive Biology of the uji fly, *Exorista sorbillans* Wied. In: *Recent Advances in uji fly Research.* (Eds) G.P. ChannaBasavanna, G. Veeranna & S.B. Dandin. pp.13-21.

64

Influence of Mating on the Temporal Oviposition Patterns of *Helicoverpa armigera* and *bombyx Mori.* L

SHOBHA, A., D. MANJULA KUMARI AND GEETHA BALI

ABSTRACT

Age specific fecundity patterns of Lepidoptera have been related to the potential importance of adult and larval nutrition in egg production. The purpose of this study was to determine the temporal oviposition patterns for two lepidopterans with different adult feeding habits and egg maturation pattern. Our results show that while *Helicoverpa armigera* females which feed as adults mated on day 4, *Bombyx mori* females which do not feed as adults began their mating activity immediately after emergence. Although the adult feeding habits and time of onset of mating were dissimilar in these insects, the temporal trends of egg laying were similar. The virgins laid eggs, although fewer in number when compared to mated moths. In mated moths, egg numbers peaked shortly after mating and rapidly declined. Temporal oviposition patterns were more erratic for unmated females, suggesting the importance of mating in establishing the shape of the oviposition curve irrespective of adult feeding habits.

65

Female Remating in Drosophila Relationship Between Frequency of Remating and Duration of Copulation

CHAYA KUMARI AND S.R. RAMESH

ABSTRACT

The female remating is an important component of reproduction in *Drosophila*. In the present study, female remating behaviour in seven subspecies of *Drosophila nasuta* subgroup viz., *D. n. nasuta, D. n. albomicans, D. n. kepulauna, D. n. kohkoa, D. s. sulfurigaster, D. s. albostrigata* and *D. s. neonasuta* were observed and compared. The frequency of female remating ranged from 29% (*D. s. neonasuta*) to 95% (*D. s. sulfurigaster*). Chi-square values showed statistically significant difference between frequency of first (virgin matings) and second matings (rematings). One-way Analysis of variance (ANOVA) revealed that the duration of copulation in first (virgin matings) and second matings (rematings) in all subspecies studied showed significant difference. Significant shorter duration of copulation was seen in first (virgin matings) as compared to second mating (remating) in *D. s. neonasuta,* while it was non significant for *D. n. nasuta* and *D. s. albostrigata*.. Significant longer duration of copulation in first (virgin matings) than second mating (rematings) was observed in *D. n. albomicans* and *D. n. kepulauna* while it was non significant in case of *D. n. kohkoa* and *D. s. sulfurigaster*. In contrast to earlier reports our study has revealed that if the first mating (virgin mating) is of longer duration, the incidence of remating is increased. It may be suggested that different subspecies in *D. nasuta* subgroup vary in the incidence of remating and duration of copulation due to differences in their reproductive biology.

66

Cold Blooded Love Behaviour in Insects and Freck Occurrence in Human Beings

MARIMUTHU, K., M. KANIMOZHI
AND PANJALY RAMARAJ

ABSTRACT

Animals display a great diversity in the patterns of association between breeding males and females. An amazing variety of complex behaviour patterns is encountered in the mating activities of various insects. Mating priority may be determined by dominance status without reference to resources and without coercion. Mating occurs when physiological maturation and given environmental conditions are effective. Mating systems represent complex constellation of characters which may very greatly within and between species. An attempt has been made to unravel mystery shrouded in the cold blood love of insects and the freck occurrence of the same in certain exceptional human beings.

67

Behavioural and Molecular Genetic Evidence of Mating Strategy in the Polygynous Mating Bat *Cynopterus sphinx*

KARUPPUDURAI, T., N. GOPUKUMAR AND K. SRIPATHI

ABSTRACT

The Indian short-nosed fruit bat, *Cynopterus sphinx,* (Megachiroptera: Pteropodidae) is a common plant-visiting bat which roosts solitarily or in small groups in the foliage. It lives in small clusters of colonies of about 2-30 individuals and is reported to exihibit a resource-based polygynous mating system. Breeding colonies typically consist of few harems with one or more solitary males nearby. Essential to understanding the mating system of *C. sphinx* is identifying the reasons for solitary roosting behaviour in adult males. In this study, we attempt to elucidate the impetus behind solitary roosting by investigating the following questions. Are trees and foliage suitable for tent-making a scarce resource? Are solitary males less competitive and so remain isolated from the breeding activities? However the role of non-harem males as probable fathers has not been examined. We carried out mark-recapture studies and using PCR based RAPD and microsatellite DNA markers to assess the paternity of harem males and nearby non-harem males to the young born in the harems. A total of 30 arbitrary primers and 9 Microsatellite DNA markers were used to identify the parentage of offspring. Our results suggest that female aggregation in *C. sphinx* cannot be attributed to scarce resource. Comparison of morphological variables between harem and non-harem males showed no significant difference. Approximately 50% of non-harem males had scrotal testes and our preliminary molecular results suggest that the non-harem males also gain access to females and sire more offspring in breeding i.e. July-August breeding seasons than February-March breeding season. These results suggest that non-harem males are reproductively active and enjoy lilmited reproductive success.

68

The Effect of Dark/Light Cycle on Mating Propensity and Productivity in *Drosophila*

JYOTHI, H.K. AND B. P. HARINI

ABSTRACT

The effect of light and dark regimes on mating propensity and fitness of *Drosophila* have been studied using lifespan as the sole index of physiological well being. It is known that mating propensity is inversely related to the reproductive output in many species. In the present study we have made an attempt to record the mating propensity and productivity (longevity, fecundity, fertility) of different species of *Drosophila*. The effects of constant exposure to light, constant exposure to darkness and alternating light and dark cycle on the mating propensity and productivity of the mated flies were observed. The results have revealed that mating propensity increased with alternating light and dark cycle rather than constant light or dark cycle and also the productivity was shown to increase with the increased mating propensity. The implications of the results will be discussed.

69

An Experimental Set Up to Study Mating in Captivity in the Primitively Eusocial Wasp, *Ropalidia marginata*

SWATI SAMUDRE, RUTUJA CHITRA TARAK
RUCHIRA SEN AND RAGHAVENDRA GADAGKAR

ABSRACT

Independent founding polistine wasps generally mate away from their nests. Males either assemble in a lek-like system or patrol marked territories and foraging areas to find mates. *Ropalidia marginata* is a primitively eusocial, independent founding polistine wasp, whose social biology has been extensively studied but mating biology has been difficult to study in the wild. To fill in this lacuna, we have attempted to study the mating behaviour in laboratory conditions and established a successful experimental set up for this purpose. This set up has allowed us to describe various behaviours associated with mating such as, attempt to mount, mount, short conjugation and long conjugation, and understand which behaviours are required for sperm transfer. We have observed 45 unique pairs of wasps between 5-20 days of age, in transparent aerated plastic boxes. Among the 45 pairs, 20 pairs successfully transferred sperm, 14 pairs attempted mating unsuccessfully and 11 pairs did not interact. We measured the durations and occurrences of different mating behaviours of the pairs who transferred sperm and of those who were unsuccessful. From these data we conclude that long conjugation is essential for sperm transfer and speculate that female choice rather than male coercion is more important for successful sperm transfer.

70

Dominant Deriving A Boost By Supressing the Subordinate: A Study on A Primitively Eusocial Wasp, *Ropalidia marginata*

VIDHI PAREEK AND RAGHAVENDRA GADAGKAR

ABSTRACT

Queens of *Ropalidia marginata* are docile as opposed to queens of other primitively eusocial wasps which are behaviourally highly dominant. In primitively eusocial insects, behavioural dominance of queen is thought to control worker reproduction and thus to maintain reproductive monopoly. In spite of being docile, a *R. marginata* queen maintains reproductive monopoly. After removing the queen, one of the individuals (Potential Queen) shows very high levels of aggression followed by monopolization of reproduction and reversion to a docile behaviour. Earlier studies indicated that a potential queen does not use her dominance to suppress reproduction by subordinate individuals. Moreover, a potential queen with nest mates develops her ovaries significantly faster than a potential queen with no nest mates, suggesting that the potential queen may use dominance for her own ovarian development. If the hypothesis is that the dominance shown by the potential queen is for her own ovarian development, the prediction would be that a female having an opportunity to show aggression should develop her ovaries faster than a solitary individual who doesn't have this opportunity. To verify the prediction, we kept virgin females either in pair or in isolation from the day of their eclosion and dissected them after 1, 2 or 4 weeks. Ovarian measurements and further analysis revealed that (1) in the case of paired females, one of the females always had more developed ovaries (reproductively dominant) than the other (reproductively subordinate) (2) the reproductively dominant female had better developed ovaries than isolated females who, in turn, had better developed ovaries than subordinate female. This observation lends support to our hypothesis and thus suggests that the increase in reproductive success of a dominant female by living in a social group is not only due to assistance in brood care but it also speeds up her ovarian development resulting in higher reproductive capability.

71

Seasonal Variation in the Size of Nasal Gland in Two Sympatric Species of Hipposiderid Bats

SWAMI DOSS, D. P., K. SRIPATHI.
AND N. GOPUKUMAR

ABSTRACT

Hipposideros speoris and *Hipposideros ater* closely resemble sympatric bat species found throughout the plains of Indian subcontinent. *H. speoris* is comparatively larger than the *H. ater* and both the male bats have prominent nasal gland. It occurs at the dorsal middle portion above the nose leaf. In females, the nasal gland is found in a rudimentary form. These bats use the nasal gland to mark their own females. The size of the gland changed in the course of annual reproductive cycle, which is characterized by peak in development during the breeding season. In *H. speoris,* the size of the gland increased from the first week of April (5.52 ± 0.28 mm length x 4.30 ± 0.67 mm breadth, n = 6) and until the last week of June and after that, there was a gradual decrease in the size of the gland. A positive linear relationship exist between nasal gland growth and testis volume ($r = 0.695$, $n = 18$, $P < 0.01$). In *H. ater* the size of the gland reached two peak values, first during April–May and second during September–October. This clearly indicates that the increase of nasal gland size is closely associated with the size of the testis and plays an important role in the breeding activities of these bats.

Recommendations of the XXXI Annual Conference of Ethological Society of India

Held at the University of Agricultural Sciences
GKVK, Bangalore-560 065
Karnataka

The XXXI Annual Conference of Ethological Society of India, held during April 10-12, 2007 at the University of Agricultural Sciences, Bangalore provided valuable glimpses of the behavioural studies under progress across the country. The behavioural scientists touched upon many important national issues requiring attention by the government and the policy makers. The scientists deciphered on tiger census, bat pollination, effects of flash floods on benthic especially fish fauna, women ethologists, impact of bamboo flowering on rodent population in the North-East, role of honey bees in crop yield improvements, herbal products for the management of Pigeon-Pea pod borers, Distance Education for knowledge society and several other issues. Experienced as well as young ethologists actively participated in the deliberations. The papers presented indicated deployment of modern tools of research like telemetry, monitoring animal populations through video cameras, neurosciences, molecular biological tools, *Drosophila* genetics, phonotactic experiments and other disciplines of modern biology. The ethologists at the conference concurred that by and large ethology is still at its infancy in India. The country being rich in biodiversity with many indigenous species is highly potential for behavioural research but still remains untapped. The discussions and interactions in the various sessions led to the following recommendations:

1. More numbers of training programmes and workshops spanning two-three days need to be conducted on Animal Behaviour across the country.
2. International collaboration, networking at college and University levels should be carried out to provide state of Art training in frontier areas of Animal Behaviour.
3. The participants lauded the efforts of the NCERT and UGC in initiating programmes to include the syllabus on Ecology and Animal Behaviour at school, Pre-University and Degree levels.
4. Nationwide creation of centres of excellence in the fields of Animal Behaviour is to be initiated.
5. Studies on behaviour in desert and marine habitats need to be stepped-up.
6. The Ethological Society of India should make all-out efforts to conduct Annual Conferences of the society in North India.
7. Studies on invertebrates and small mammals need to be taken up as they are neglected. Most of the time researchers have been concentrating only on flagship species.
8. Studies on the behaviour of invasive species are crucially important to conserve native species and should be initiated.

Subject Index

A

F

G

H

I

K

L

M

N

O

P

Q

R

S